The Imperial College Lectures in
PETROLEUM ENGINEERING

Drilling and Reservoir Appraisal

Volume
4

The Imperial College Lectures in
PETROLEUM ENGINEERING

Drilling and Reservoir Appraisal

Volume
4

Olivier Allain
KAPPA, France

Michael Dyson
Striatum Ltd, UK

Xudong Jing
Shell, Netherlands

Christopher Pentland
Petroleum Development Oman, Oman

Marcel Polikar
Independent Consultant, Canada

Vural Sander Suicmez
Maersk Oil & Gas, Denmark

World Scientific

NEW JERSEY · LONDON · SINGAPORE · BEIJING · SHANGHAI · HONG KONG · TAIPEI · CHENNAI · TOKYO

Published by

World Scientific Publishing Europe Ltd.

57 Shelton Street, Covent Garden, London WC2H 9HE

Head office: 5 Toh Tuck Link, Singapore 596224

USA office: 27 Warren Street, Suite 401-402, Hackensack, NJ 07601

Library of Congress Cataloging-in-Publication Data
Names: Allain, Olivier, editor. | Dyson, M. R. (Michael Richard), 1964– author. |
    Jing, Xudong, author.
Title: Drilling and reservoir appraisal / (author) Olivier Allain (Kappa, France), Michael Dyson
    (Striatum Ltd, UK), Xudong Jing (Shell, Netherlands), Christopher Pentland
    (Petroleum Development Oman, Oman), Marcel Polikar (Independent Consultant, Canada),
    Vural Sander Suicmez (Maersk Oil & Gas, Denmark)
Description: [Hackensack] New Jersey : World Scientific, 2017. | Series: The Imperial College
    lectures in petroleum engineering ; volume 4 | Includes index.
Identifiers: LCCN 2017024335 | ISBN 9781786343956 (hardcover : alk. paper)
Subjects: LCSH: Oil well drilling. | Oil reservoir engineering. | Oil wells--Measurement. |
    Petroleum reserves. | Petroleum--Geology.
Classification: LCC TN871.2 .D7284 2017 | DDC 622/.3381--dc23
LC record available at https://lccn.loc.gov/2017024335

British Library Cataloguing-in-Publication Data
A catalogue record for this book is available from the British Library.

Disclaimer
All reasonable effort has been taken to obtain permission for extracts, figures and images within this book. If there are any examples where you believe copyright has been infringed, please contact Imperial College London with details of the content in question.

For any available supplementary material, please visit
http://www.worldscientific.com/worldscibooks/10.1142/Q0115#t=suppl

Desk Editors: Anthony Alexander/Jennifer Brough/Shi Ying Koe

Typeset by Stallion Press
Email: enquiries@stallionpress.com

Foreword

The oil industry has long been accustomed to fluctuations in the price of the oil, volatility in geopolitical landscape and far reaching new developments in technology and innovation. This forces the oil industry, including operators, service providers and suppliers as a whole, to adapt to such environments by cutting costs and reducing uncertainties in making critical investment decisions. In order to make such changes and adapt to a variable price environment that can create low profit margins, the industry is forced to make continuous technical improvements, especially in the context of reducing the cost of enabling technologies.

Technical advances have been a true game changer in oil industry. The successful application of new technologies has literally enabled oil companies to transform resources which were once thought unconventional into conventional ones. As the saying goes "what is impossible today will be easy tomorrow". Drilling technology is one such technology which has made some radical changes in oil industry in the recent years that impacts the world's macroeconomic outlook and the geopolitical future.

This volume (Volume 4: *Drilling and Reservoir Appraisal*) covers the fundamentals and recent advancements in drilling technology and reservoir appraisal strategies. It is a unique tribute to practising engineers and geoscientists in the field as well as to the students studying in petroleum engineering and earth science programs. This is an important part of the book titled *The Imperial College Lectures in*

Petroleum Engineering since a careful well design and an appropriate reservoir appraisal program are the first steps towards a successful development of an oil field while staying within the predefined economic bounds and minimising the risks and uncertainties in relevant applications.

The authors are experts in their own field with the right blend of both industrial and academic experience. Therefore, this book offers a unique opportunities for the readers to learn both theory and practice from academic to industrial perspectives. It is without hesitation that I recommend earth science and petroleum engineering professionals to have this book in their collection as I am confident that they will find it a useful reference book in the specific area of "Drilling and Reservoir Appraisal".

Dr. Birol Dindoruk

Chief Scientist, Reservoir Physics
Shell International E&P Inc.

Preface

As part of the *Imperial College Lectures in Petroleum Engineering*, and based on a lecture series on the same topic, *Drilling and Reservoir Appraisal* provides the introductory information needed for students of earth sciences, petroleum engineering, engineering and geoscience. The authors of this book are renowned experts in the industry specializing in the very areas they write about.

This book is organized into three chapters.

In Chapter 1, with many real-life examples sandwiched between the topics, Michael Dyson first introduces the basic principles of well engineering, in terms of planning, design and construction, and its impact on optimal field development. He then moves on to look at the basic drilling and completion stages and equipment used in the process, laying the ground for students to appreciate well operations safety, costs and operations management. Michael Dyson has been working in the oil and gas arena for many years all across the globe for corporations like Shell, BG Group and Navigant.

Chapter 2 introduces the concepts of core analysis focuses on issues of coring and the laboratory measurement of the physico-chemical properties of samples, and underlines the importance of hydrocarbon reservoir development. This chapter is written by Vural Sander Suicmez, Marcel Polikar, Xudong Jing and Christopher Pentland who are all technical experts in petroleum engineering with decades of industry and academic experience.

Known as petrophysics in the context of laboratory measurements of core material from petroleum reservoirs, core analysis remains an important element within this domain to examine the physico-chemical properties of samples of recovered core for purposes within multiple disciplines. Petrophysics also encompass well log data acquisition and interpretation.

Chapter 3 focuses on Production Logging (PL), an in-well logging operation designed to describe the nature and the behaviour of fluids in or around the borehole, during either production or injection. PL provides, at a given time, phase by phase and zone by zone, how much fluid is coming out of or going into the formation. To obtain these information, the service company engineer will run a string of dedicated tools to capture and process the information. PL may be run for different purposes: monitoring and controlling the reservoir, analysing dynamic well performance, assessing the productivity or injectivity of individual zones, diagnosing well problems and monitoring the results of a well operation (stimulation, completion, etc.). In some companies, the definition of PL extends up to what we call cased hole logging, including other logs such as Cement Bond Logs (CBL), pulse Neutron Capture Logs (PNL), Carbon/Oxygen Logs (C/O), corrosion logs, radioactive tracer logs and noise logs. Olivier Allain, technical director and petroleum engineer at KAPPA, explains the main methods to interpret classical and multiple probe production logs in this segment.

About the Authors

Olivier Allain is Technical Director of KAPPA, a petroleum engineering software company based in Sophia-Antipolis, France. During his 27 years with KAPPA, he has been involved with the development of interpretation software in disciplines ranging from pressure and rate transient analysis to production logging.

Michael Dyson is a Director of Striatum, a consulting company delivering projects in the oil and gas arena. Over the course of his career, he has worked worldwide at Shell, BG Group and Navigant Consulting. He is also an Industry Guest Lecturer at Imperial College London.

Xudong Jing is the General Manager of improved and enhanced oil recovery technology at Shell. He has held petroleum engineering technical and leadership positions in the UK, Oman, China, US and the Netherlands. He is also a Visiting Professor in petroleum engineering at Imperial College London.

Christopher Pentland is a Reservoir Engineer at Petroleum Development Oman. He has previously worked for Shell Global Solutions and studied at the Department of Earth Science and Engineering at Imperial College London.

Marcel Polikar is an Independent Consultant based in Canada. He has worked as a Professor, Coach and Mentor for the University of Alberta and as a Principal Reservoir Engineer for Shell International.

Vural Sander Suicmez is a Lead Reservoir Engineer at Maersk Oil and Gas in Copenhagen, Denmark. Before joining Maersk, he has held positions with Shell in the Netherlands and Brunei and with Saudi Aramco in Dhahran, Saudi Arabia. He is also a Visiting Lecturer in petroleum engineering at Imperial College London.

Contents

Chapter 2. Core Analysis 231

Vural Sander Suicmez, Marcel Polikar, Xudong Jing and Christopher Pentland

Chapter 3. Production Logging   301

Olivier Allain

Chapter 1

Well Engineering

Michael Dyson

Director, Striatum Limited, UK

This chapter of the Handbook of Petroleum Engineering should enable the reader to:

- articulate the basic principles of well planning, design and construction;
- understand how well design and construction contributes to optimal field development;
- recognise the basic drilling and completion stages and equipment used;
- possess a basic understanding of alternative completion designs;
- appreciate well operations safety, costs and operations management.

1.1. Introduction

1.1.1. *Objectives*

Wells play a key role in every phase of the field development from exploration, through appraisal, development, the production phase and field abandonment. In a typical deepwater offshore project, well costs can account for 50% of total project Capital Costs (CAPEX). The quality of information acquired from an exploration or appraisal well can determine the efficacy of a multi-billion dollar investment. Avoiding formation damage during the drilling process can make a substantial difference to the well's production. Placing the well optimally in the reservoir strongly affects field drainage and ultimate recovery. And planning for future requirements and overall well integrity maximises its business contribution over the full lifecycle of the field.

Proper management of drilling operations is essential, given that operating costs may exceed $1 million per day for a deepwater rig. The need for diligent supervision and operations has been exemplified by blowouts of wells such as that on BP's Macondo field in the US Gulf of Mexico.

All these aspects point to the value of applying project management concepts to well delivery. This implies early involvement in field development concepts and application of a Well Delivery Process as described later in this chapter.

The exploration and production business generally, and drilling in particular, is rife with terminology, abbreviations, three-letter acronyms (and imperial units). When first used, specific terms are noted in **bold**. A glossary is provided at the end of this chapter.

1.2. Health, Safety and Environment (HSE)

Well engineering operations such as drilling, completing and testing wells are potentially very hazardous activities. A penetration is being made into deep, naturally pressured formations often containing flammable liquids and gases that may also be poisonous. If these flow uncontrolled to surface and catch fire, the result can be catastrophic — with likely human fatalities and injuries, severe environmental damage, loss of equipment and impact on the reputation of the companies and organisations involved.

Furthermore, all drilling activities involve using and moving heavy equipment around on the wellsite, with the potential to injure those working there. High-powered machinery, caustic chemicals and explosives are frequently used in drilling operations.

Rig operations have tended to be a highly cyclical business, with "booms and busts" resulting in activities starting and stopping at short notice. Historically, this in turn has resulted in unskilled labour being recruited at the wellsite, particularly for land operations, with a high injury rate being the almost inevitable outcome. There are old jokes — not far from the truth — that "real" drillers have a few fingers missing from past injuries.

The industry has addressed and continues to address these challenges, for example, by hiring and rigorous training of drilling crew, and making increasing use of automated systems on the rig-floor. Government regulators and agencies have demanded improvements in safety, healthy working conditions and considerably reduced environmental impact from well engineering activities. As a result, the Exploration & Production (E&P) business, and rig operations in particular, is very HSE-conscious and enjoys occupational injury rates lower than most other industries.

In line with the philosophy that HSE permeates all aspects of well operations, this chapter will identify and discuss specific HSE considerations as they arise. In the meantime, a few key concepts need to be covered here.

1.2.1. *Accidents are not Accidental*

All accidents are incidents that arise from a series of failures. These are sometimes shown in the "Swiss Cheese model" after James Reason — the concept being that the barriers erected to prevent an accident tend to have holes in them and that an accident will happen if all the holes in all the barriers line-up (see Figure 1.1).

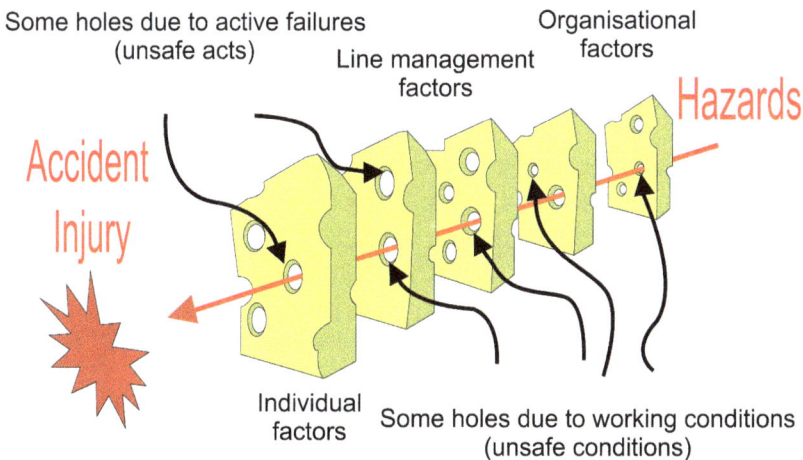

Figure 1.1. Swiss Cheese safety model.

This model is used in many industries including the airline business and medical practice. The barriers may be physical (such as a hard-hat), avoiding unsafe acts, following rules and procedures, possessing competence and effective supervision.

The industry focuses on "plugging the holes" in the barriers, testing them to make sure they are reliable and reporting any shortfalls.

Additional measures are also taken to mitigate the results of an incident (as opposed to avoiding the incident in the first place). These include emergency response capability, fire-fighting equipment and first-aid training.

1.2.2. *Safety is Built-In*

It can be appreciated that many of the barriers to accidents and incidents result from:

- proper planning of operations, automating and removing people away from high risk tasks if possible;
- providing appropriate equipment including PPE (personal protection such as hard-hats, boots, eyewear, coveralls and gloves);
- competent, skilled and motivated workforce;
- strong leadership and decision making;
- a culture of reporting incidents, learning from them and implementing improvements.

1.2.3. *Personal Safety*

What does this mean for a typical petroleum engineer in the E&P business? It is essential to have an awareness and appreciation of the risks to personally stay safe and to watch out for the safety of others. In responsible organisations, this is a value required of all staff — to take HSE seriously and:

- if in doubt STOP the operations;
- ask questions. Well construction is a collaborative business;
- share information and knowledge;

- never assume the other person(s) know(s);
- always ask yourself "What if?";
- continuously check and recheck;
- know the barriers that are keeping you safe.

1.3. Well Delivery Process

As mentioned above, well engineering plays a vital role in the overall profitability of an oil or gas development project.

Figure 1.2 shows the notional project cashflow of such a project over its lifetime — here assumed to be 30 years. Typically, CAPEX investments are made in the early years in production equipment and wells and subsequently in side-tracking wells to more remote

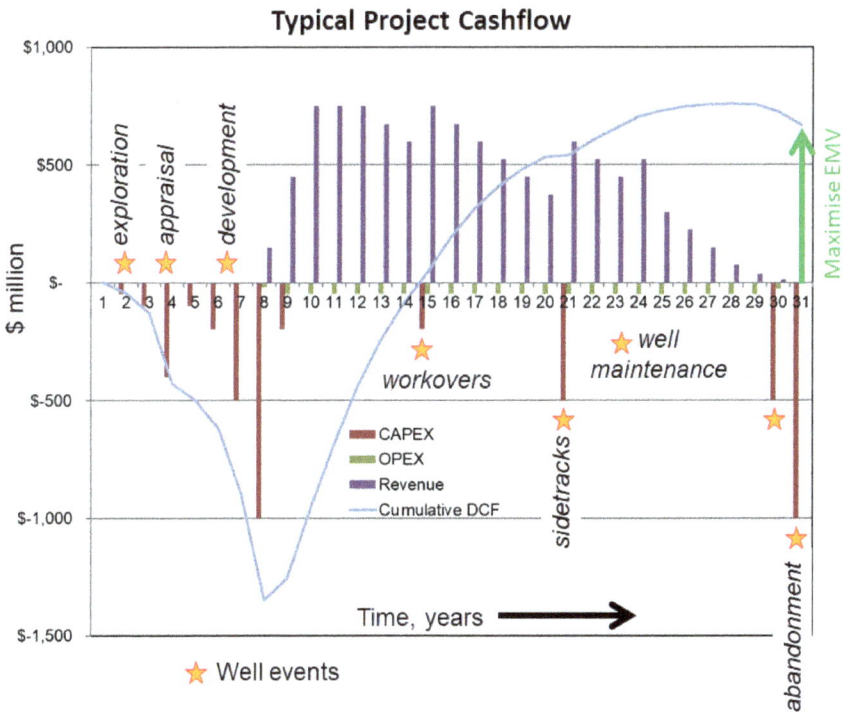

Figure 1.2. Typical E&P project cashflow.

areas of the field. Operating Costs (OPEX) occur throughout the life cycle arising from operating the facilities and maintaining the wells. These costs are offset by the revenue from production. The net result provides an overall Discounted Cash Flow (DCF) and results in the Estimated Monetary Value (EMV) of the project. During the project, the following factors can be influenced to maximise project profitability:

- maximising hydrocarbon recovery;
- optimum data gathering;
- manage risks;
- meet or exceed HSSE standards;
- right competencies — technical, commercial;
- suitable partner selection;
- stakeholder management;
- appropriate technology and innovation;
- multi-disciplinary teams;
- clear business processes and management of change;
- clear accountabilities and leadership.

The following table describes the main well engineering activities in each of the project phases:

Project phase	Typical well activities	Focus
Exploration	Drilling exploration well based on seismic interpretation of the prospect	Physical confirmation of hydrocarbons
Appraisal	Drilling and production test of one or more appraisal wells	Confirm areal extent of field, test fluid mobility and properties
Development	Drilling and completion of production wells	Lowest cost, efficient operations, optimum placement

(*Continued*)

(*Continued*)

Project phase	Typical well activities	Focus
Operations	Maintenance and repair of wells. Recompletion, stimulation and/or side-tracking of wells	Well up-time, future operability and getting the most out of the asset
Abandonment	Well re-entry, plugging and abandonment	Minimal cost to achieve effective abandonment (and reclamation if on land)

1.3.1. *Well Delivery Process*

A well-defined and documented well delivery process can be useful to improve the quality of the wells, and hence overall business performance. This is achieved by:

- identifying responsible parties for key decisions;
- promoting multi-disciplinary teamwork;
- driving appropriate risk management;
- facilitating introduction of new ideas/approaches;
- setting demanding targets and challenging the organisation to meet them.

Most E&P operators have documented their well delivery process into the five stages as shown in Figure 1.3. The closed-loop demonstrates the learning that should take place from the overall process (there are also learning loops within each phase and between different projects following the process). The process is applied to a single well — often the case for exploration — or for a series of wells if they are similar, as in the case of production wells. Well(s) will progress from one stage to the next through a "decision gate" in which management and technical experts will review the proposal

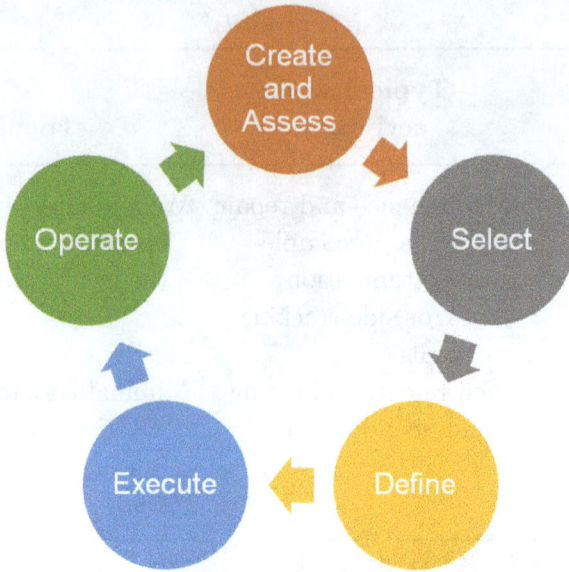

Figure 1.3. High level well delivery process.

and signoff that all relevant work has been done and that the project is still valuable and planned for execution.

In each of these five phases the following takes place:

Phase	Typical duration (months)	Activities
Create & Assess	2–12	• Initiate project (an opportunity) • Prepare design concepts • Identify field (or well) concept options • Review field (or well) concept options

(*Continued*)

(Continued)

Phase	Typical duration (months)	Activities
		• Identify long lead items, e.g. special rigs, special equipment, CRA materials • Run provisional economics

Select	2–12	• Evaluate feasibility
		• Select best option
		• Confirm the well design meets objectives:
		  ◦ Information
		  ◦ Production rate
		  ◦ Life cycle
		• Confirm economic justification

Define	4–8	• Complete detailed well design • Carry out a peer review • Develop, optimise and finalise well programme • Complete execution plan
Execute	1–12	• Obtain approval for detailed well design • Obtain approval for expenditure • Conduct drill well on paper exercise • Drill/complete/workover the well • Review performance
Operate	0–600	• Evaluate operational performance • Complete end of well report • Catalogue learnings; transfer to next well in sequence • Share learnings

1.3.2. *Drilling Programme*

An important deliverable from the define stage is the drilling programme, and it is normal to produce this specifically for each well to be drilled. This document describes how the well will be drilled in technical detail, and many of the elements will be addressed in this chapter. The front sheet of typical drilling programme covers the essential elements and is shown in Figure 1.4.

Most international oil and gas companies contract out the drilling operations to a Drilling contractor, who provides and operates the drilling rig following the instructions from the operating company who takes responsibility for the overall activity. Other, specialist activities are also contracted by the operator. These companies typically provide the following services:

- directional drilling/LWD/surveying;
- drilling fluids;
- cementing;
- petrophysical logging;
- mud logging;
- casing/tubing running;
- cuttings management;
- other equipment specialists;
- well testing;
- inspection;
- logistics (boats, trucks, helicopters);
- site construction;
- communications;
- accommodation and catering;
- lifting;
- environmental management;
- local liaison;
- security.

Contracts for these services need to be organised and in-place before drilling the well.

Hole size and depth | **Casing** | **Cement** | **Drilling Fluid** | **Deviation** | **Logging**

Well name, rig name, budget & time, surface & target coordinates

| E&P Company Drilling Programme Participation: Shell 45% ABC 30% EnerCo 25% | Surface coordinates: 12deg 34min 56sec W 56deg 34min 12sec N Transverse Mercator: 6365214 m N 324 410 m E Seismic line XS.911.53.1 | Rig: Leachim Nosyd III Type: 10K Semisubmersible, chain moored Contractor: Neb Namiook Estimated spud date: 1/1/97 | Well name: Recruit-1 Well Type: Exploration Confidential? - NO Block: UK-1 Water depth: 1200m DFE: 12m above sea level | Objective: Exploration of the seismic bright spots located between M-4 and M-5 markers Targets (Geological tolerances): i) +/- 200 m square at M-4 marker ii)+/- 200m square at M-5 marker Note: Vertical tolerance zero. | Work order: A12345B/1 Objective: Drill, evaluate and test if HC observed on logs Cost: $18.38 million (dry hole basis) Duration: 108 days |

Expected geology

30" cond 1270m
20" csg
13 3/8" csg 2200m
9 5/8" csg 3800m
7" liner 3800m
Fault

Hole size and depths m ahbdf (m hvss)	Casing set m at ahbdf (m hvss)	Cement Type	TOC m TVD	Drilling Fluid Spec	Density kPa/m	Deviation Surveys (frequency)	Evaluation Logging	Pressure Tests kPa BOP	Casing	Geology markers
36" (combination 12 ¼" bit and 36" hole opener run together) 1272 m ahbdf (1260 m hvss) seabed at 1212 (1200)	30"X-56/1.5"WT/RL-4 1270 m ahbdf (1258 m hvss)	Class "G" Seawater 1.90 SG 200% excess	Seabed	Seawater high-vis pills	None	MSS (150m)	none	Test on stump 69000/ 34500 kPa	None	
26" 1505 m ahbdf (1493 m hvss)	20"/9/40.625" WT/J-55/RL-4s 1500 m ahbdf (1488 m hvss)	Class "G" Freshwater Lead 1.62SG Tail 1.90 SG 150% excess	Seabed 1300 m	Gyp. CMC high YP, PV	10 kP/alm	MSS (at seabed and TD)	None	34500 10500 (Ann)	6900 (After WOC)	M-1 marker 2210 m TVD Top closure 2511 m TVD
17 ½" 2205 m ahbdf (2193 m hvss)	13 3/8"/54/J55/BTC 2200 m ahbdf (2188 m hvss)	Class "G" Freshwater Lead 1.62SG Tail 1.90 SG no excess	1600 m 2000 m	Gyp CMC high YP, PV	11 kP/alm	MWD MMS (TD to bump)	None	34500 24100 (Ann)	20700 (at bump)	M-2 marker (top salt) 2210 m TVD M-3 marker (base salt) 2210 m TVD
12 ¼" 3805 m ahbdf (3444 m hvss)	9 5/8"/47/N80/BTC down to 1970 m ahbdf then 9 5/8"/40/N80/BTC 3800 m ahbdf (3441 m hvss)	Retarded Class "G" Freshwater Lead 1.62SG Tail 1.90 SG caliper +20%	Seabed 3200 m	Gyp CMC high YP, PV FL uncontrolled	13 kP/alm	MWD MMS (TD to shoe)	SGR/DLL/MSFL/DSI SGR/DL/CN/CAL MDT (possible) CST s(60)	51700 24100 (Ann)	20700 (at bump)	Depleted sand stringers 3000m TVD F-1 Fault 3400m TVD
8 ½" 5002 m ahbdf (4213 m hvss)	7"/23/N80/BTC 5000 m ahbdf (4212 m hvss)	Retarded Class "G" with F.L Control Freshwater 1.90 SG caliper	TOL 3650 m ahbdf	KCl/M, PV=10, YP=10	17 kP/alm	MWD MMS (TD to shoe)	SGR/DLL/MSFL/DSI SGR/DL/CN/CAL MDT (+ samples) CST s(60) Offset VSP	51700 24100 (Ann)	20700 (at bump)	M-4 marker (top sand) 3500m TVD M-5 marker (base sand) 3650m TVD

Approval Signatures

| Written by: | ABC/11 | ABC/1 | ABC | AB | AB/D | Exploration Manager |
| Approved by: | TAB/11 | TAB | TA | XYZ | | Operations Manager |

One row for each hole section

Figure 1.4. Example drilling programme summary page.

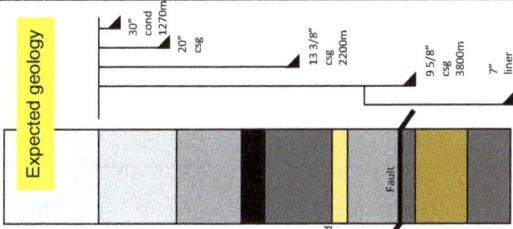

1.3.3.   *Overall Sequence of Drilling Operations*

Physically, well engineering encompasses more than drilling wells — it includes the following activities that may be stand-alone or combined with others as necessary:

Operation	Description
0 Drive conductor	Pile-drive the top well casing
1 Spud	Start a new well
2 Trip pipe	Run Drill pipe (or other tubular) into or out of the hole
3 Make-Up/Break-down Bottom-Hole Assembly (BHA)	Screw together/unscrewing BHA tubulars
4 Drill	Actually cutting rock to make a hole
5 Circulate	Pump fluids around (into and out) of the well
6 Steer	Drill the well in a certain direction
7 Log	Obtain petrophysical or other data from the well
8 Survey	Measure the trajectory of the well
9 Run casing	Lower well casing into the ground
10 Cement casing	Pump cement around the casing to seal it and fix it into the ground
11 Pressure test	Check the physical integrity of the well
12 Formation integrity test	Check the physical integrity of the rock around the well

(Continued)

(*Continued*)

Operation	Description
13 Nipple-up/down BlowOut Preventers (BOPs)	Connect/disconnect BOPs to/from the well
14 Run tubing	Lower production tubing into the well
15 Nipple-up/down Christmas tree	Connect (disconnect) Christmas tree to/from the well
16 Run wireline tools in/out of the well	Lower into and recover from the well tools for electric logging
17 Run slick-line tools in/out of the well	Lower into and recover from the well tools for setting plugs, shifting sleeves etc
18 Perforate	Make holes in the casing to allow hydrocarbon flow
19 Stimulate	Pump fluids into the formation to increase productivity
20 Gravel pack	Place gravel in the well to stop sand production
21 Produce	Operate the well to produce hydrocarbons
22 Inject	Inject fluids into the well
23 Analyse	Analyse data from the well and use to optimise production and maintain the well
24 Maintain	Maintain the well in safe working condition
25 Workover	Reconfigure the well for future production

(*Continued*)

(Continued)

Operation	Description
26 Sidetrack	Drill from an existing hole to a different location
27 Suspend	Leave the well in a safe non-producing condition
28 Abandon	Close-up the well permanently

Note: Steps 2–13 recur several times until the well reaches the required depth.

All these activities are described and explained in more detail later in this chapter.

1.4. Regulations and Standards

Well engineering is covered by a number of regulatory requirements and other standards, such as:

- governmental/regulatory standards — HSE;
- company policies and standards;
- industry standards;
- third-party standards, e.g. drilling programme review, inspection and rig insurance.

Government (federal, state, EU) standards typically require "good oilfield practice" in well engineering operations. Sometimes this is not well defined and has been clarified by case law. Other requirements are much more specific — for example, the requirement to provide copies of wellwork programmes, the regular testing of well control equipment, keeping of operational records, accurate surveying of wellbore locations and notification of safety or environmental incidents. In some jurisdictions, active consent is required before proceeding with certain operations like abandoning a well. Regulations also cover required competencies for individuals in key roles at the wellsite — possession of a current International Well Control Certificate (IWCF) for example, or a Master's Ticket for seamanship roles. General industry safety requirements apply, and there is increased scrutiny of any environmental discharges. Government bodies usually have the right to visit and inspect rig-sites at any time and can shutdown operations if not satisfied that requirements are being met. Prosecution of companies and individuals is not unknown.

Operating company, drilling contractor and service companies typically have their own policies and standards that must be upheld, and guidelines advising best practices on how to do so. These might typically cover how to test BOPs, what drilling fluid tests are required, the detailed drilling programme, various commercial contracts, competencies that are required in certain roles, equipment that must be on the rig and many, many other elements.

There are several industry bodies — national and international — that are relevant to well engineering activities that set industry standards which are usually considered to be the minimum acceptable. These include:

- American Petroleum Industry (API) — particularly for technical equipment specifications;
- Society of Petroleum Engineers (SPE);
- International Association of Drilling Contractors (IADC);
- Oil and Gas UK.

It is usual (particularly for smaller companies) to have key activity programmes (such as a drilling programme) reviewed by an external independent party to identify any safety or operational concerns. For insurance purposes, rigs also need to be inspected and in some cases the location approved before operations can commence. A number of independent organisations (for example Det Norske Veritas (DNV)) manage these standards.

All the above must be met and where there are conflicting requirements, the most exacting must be followed. The Company Man and Drilling Contractor Toolpusher, in particular, must be familiar with all these requirements in detail.

1.5. Onshore Rigs

1.5.1. *Basic Drilling Rig*

The basic drilling rig is required to carry out the following four functions:

- hoist and lower the drill string;
- rotate the drill string;
- circulate liquid down the drill string and back up the hole;
- control pressures.

These functions will each be addressed in detail later in this chapter. Suffice to say that the technical requirements of the rig will be determined from specifying the hole to be drilled.

The type of drilling rig is determined by:

- geographical location;
- environment;
- depth of well;
- type of well;
- mobility requirements;
- operating cost.

The following generic types of rigs will be covered in this section:

- land;
- barge;
- platform;
- gravity base;
- jack-up;
- semisubmersible;
- tension leg;
- spar;
- deepwater drillship;
- coiled tubing;
- snubbing unit.

1.5.2. *Masts and Derricks*

Masts and derricks are rather similar as shown in Figures 1.5 and 1.6. In plan section, the wellbore centre lies inside the derrick, whereas, it lies to one side of the mast. Consequently, a derrick provides a stronger structure for operations and can be designed to lift a load of up to 1000 MT. A land rig operating in the desert is shown in Figure 1.7 — this also shows how a flat area (called a well pad) has been created in the local environment to site the rig and support equipment and materials for the drilling activities.

An important consideration for land rigs is that they can be rigged-up and -down, and moved to the next drilling location efficiently and safely. In cases where a well-pad accommodates several wells, this is simply a matter of sliding ("skidding") the rig to a new well while not moving much of the other equipment on the wellsite.

Figure 1.5. Drilling rig derrick.

Figure 1.6. Drilling rig mast.

Several techniques for moving rigs quickly have been developed — Figure 1.8 shows a mast-based rig on the move in a desert environment. Some drilling rigs are built onto the back of trucks to simplify moving between well locations.

Another variation is the heli-rig, which is designed for operations (often exploration wells) in very remote areas where building a road to transport the rig into location is uneconomic or impossible (see Figure 1.9). These are very light rigs that can be lifted in parts by helicopter and reconstructed on-site.

Some land rigs such as those shown in Figure 1.10 are designed for arctic operations and hence are "winterised" for low temperatures and cold winds.

1.5.3. *Highly Automated Land Rig*

A new design of land rig — for example, typically used for drilling shallow tight shale gas or coal-bed methane wells — uses new

Figure 1.7. Desert drilling operations.

Figure 1.8. Moving a light land rig by road.

Figure 1.9. Heli-rig supply in jungle location.

technology for optimising the process. Such wells are quick to drill, perhaps taking only 2–3 days, so time to move the rig becomes very significant. Hence, these rigs can be moved quickly, breaking down into small discrete truck-sized modules that can be transported over normal country roads (for example in Louisiana, and Pennsylvania, USA or Queensland, Australia). The modules are designed to be easily connected-up on site. The mast may be a single vertical beam, with the hook suspended from a hydraulic ram or raised and lowered with a rack-and-pinion arrangement rather than being conventionally strung. Both options allow pipes to be pushed into the well. Individual 40 ft drill pipes are used rather than 2–3 pipe stands. These rigs require very small pads for operation — reducing the costs and environmental impact. A noise-deadening surround can be used to minimise impact on local residents or wildlife (see Figure 1.11).

Figure 1.10. Winterised land rig.

1.5.4. *Swamp Barge Rigs*

These drillings units are designed for drilling wells in very shallow water in inland swampy environments, for example, in Nigeria — see (Figure 1.12). They are typically floated between locations and are then ballasted down onto the swamp floor prior to drilling which then proceeds in a similar way to drilling a land well. The barges are not designed for rough water and are pinned in place with legs or posts that extend down to the swamp floor. Well locations are prepared by dredging the mooring site and any access channels needed.

In water depths greater than a few metres, or where, for example, an artificial island cannot be constructed to accommodate a drilling site, more sophisticated structures are required.

Figure 1.11. Automated land rig.

Figure 1.12. Swamp barge.

1.6. Offshore Operations

For offshore operations, there are many different approaches to support the drilling rig at the surface. Depicted below, the options may be summarised in the following table:

Types of offshore platform

Type	Comments	Water depth, ft
Fixed platform steel jacket	Low cost, the only option for shallow offshore	10–1,500
Fixed platform concrete jacket	Low cost, potential to include storage cells	200–1,500
Compliant tower		1,500–3,000
Floating production system	Subsea wells. Potentially reusable	100–10,000
Tension leg platform		500–7,000
Subsea system	Requires mobile drilling rig for well operations. Needs to be tied back to land or another platform type (maximum offset depends on fluid type, pressures etc.)	100–7,000
Spar	Currently this technology offers access to the deepest water (e.g. Shell Perdido in the US GoM)	500–10,000

Several of these types of offshore platforms are illustrated in Figure 1.13.

Types of offshore drilling rig

Type	Comments	Water depth, ft
Platform rig	See above	As above
Tender assist	See above	As above
Jackup	Through jacket or standalone (e.g. exploration or appraisal) wells	0–500
Semi-submersible	Subsea wells. Stable in poor weather/sea state	50–10,000
Drillship	Subsea wells. Fast moving between locations	100–10,000
Floating Production, Drilling, Storage and Offloading (FPDSO) vessel	Subsea wells. Potentially reusable	100–10,000

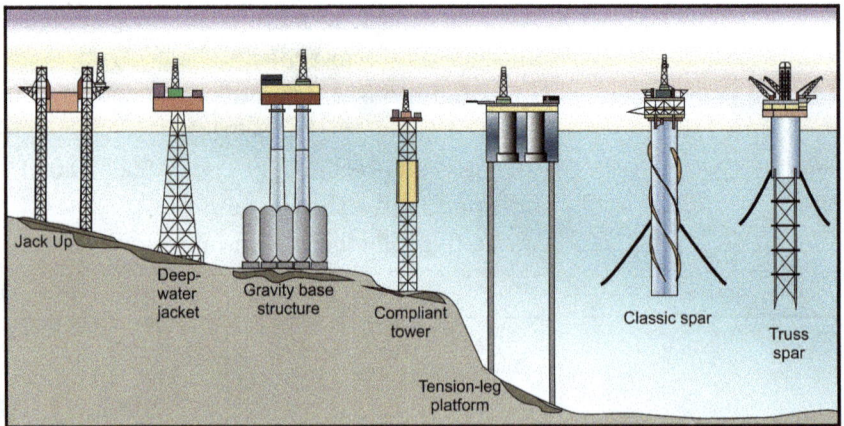

Figure 1.13. Development of deepwater installations.

1.6.1. *Steel Jacket Fixed Platform*

Common parlance has tended to confuse the terms offshore platform and drilling rig. An offshore platform comprises a support structure plus drilling rig, processing facilities, power generators, offices and living quarters. Hence the drilling rig is only one part of the overall platform and is in many ways very similar to a land rig. Some examples are shown in Figures 1.14 and 1.15.

These platforms are supported by a steel jacket that sits on the seabed and is piled so that it cannot move after final positioning. The jacket is typically lifted and lowered onto the seabed using a heavy-lift barge or alternatively floated onto location and deballasted

Figure 1.14. Schematic of steel jacket installations.

Figure 1.15. Steel jacket platform.

in a controlled way until it settles on the seabed. Water depths for this type of jacket can extend from just a few metres to 500 m or so, beyond which other designs become more economic.

As shown, the drilling rig sits on the jacket complete with all the equipment required to drill the well. There are cranes and deck space for the storage of tubulars, well equipment and drilling fluid chemicals, as well as tanks for drilling fluids. Also on the jacket are process equipment, power generation plant, office facilities, living quarters, lifeboats and a helideck. The flare is used to burn off small quantities of gas in the event of a processing interruption. The oil and/or gas produced from the wells is processed on the platform and then exported down a pipeline laid along the seabed to shore or to another platform or loading facility. The equipment sitting on the jacket is collectively called the "topsides".

The wells on a platform are drilled through conductor pipes typically 30″ in diameter that extend from the drilling rig to a depth of about 50 m into the seabed, which can be drilled or hammered into position like foundation piles. Between one and 60 wells may be drilled from a single platform, and at surface they are typically 2–3 m apart. The rig is moved from one well to another by "skidding" along supporting beams along and across the platform.

All platforms and mobile offshore drilling units are supplied by supply boat. People and smaller requirements are transported by helicopter. Cranes are provided on the installations for unloading boats.

Many types of rigs described in this chapter are mobile in that they can be moved from one well to another even if separated by many kilometres. However, platforms remain on one location for the field life — often 30 or more years. Hence, platforms are used only for development wells, i.e. once the field has been discovered. Occasionally, appraisal wells are drilled from platforms to delineate the outer reaches of a field that is already under production.

1.6.2. *Gravity Base Platform*

An alternative to the steel jacket is the so-called gravity-base which relies on its own weight resting on the seabed to support the topside equipment, which is essentially the same as for the steel-jacket platform. Some examples are shown in Figure 1.16. In Figures 1.17 and 1.18, a single-leg concrete platform is being towed to final location from the construction yard where jacket and topsides have already been mated. In this mode, the jacket is partially ballasted down and will be fully ballasted down once on location. In this case, the well conductors run inside the concrete jacket leg(s). The legs are hollow and designed to withstand seawater pressure from the outside. These structures are "slip formed" — a technique developed by contractors in fjord locations involving pouring concrete into a mould that moves up the structure which is simultaneously deballasted. In some cases, oil can be kept in storage cells at the base of the legs, also of concrete construction.

1.6.3. *Tension Leg Platforms (TLPs)*

A third type of platform, the TLP is essentially a floating structure that is linked to the seabed via steel tendons. The tendons are piled

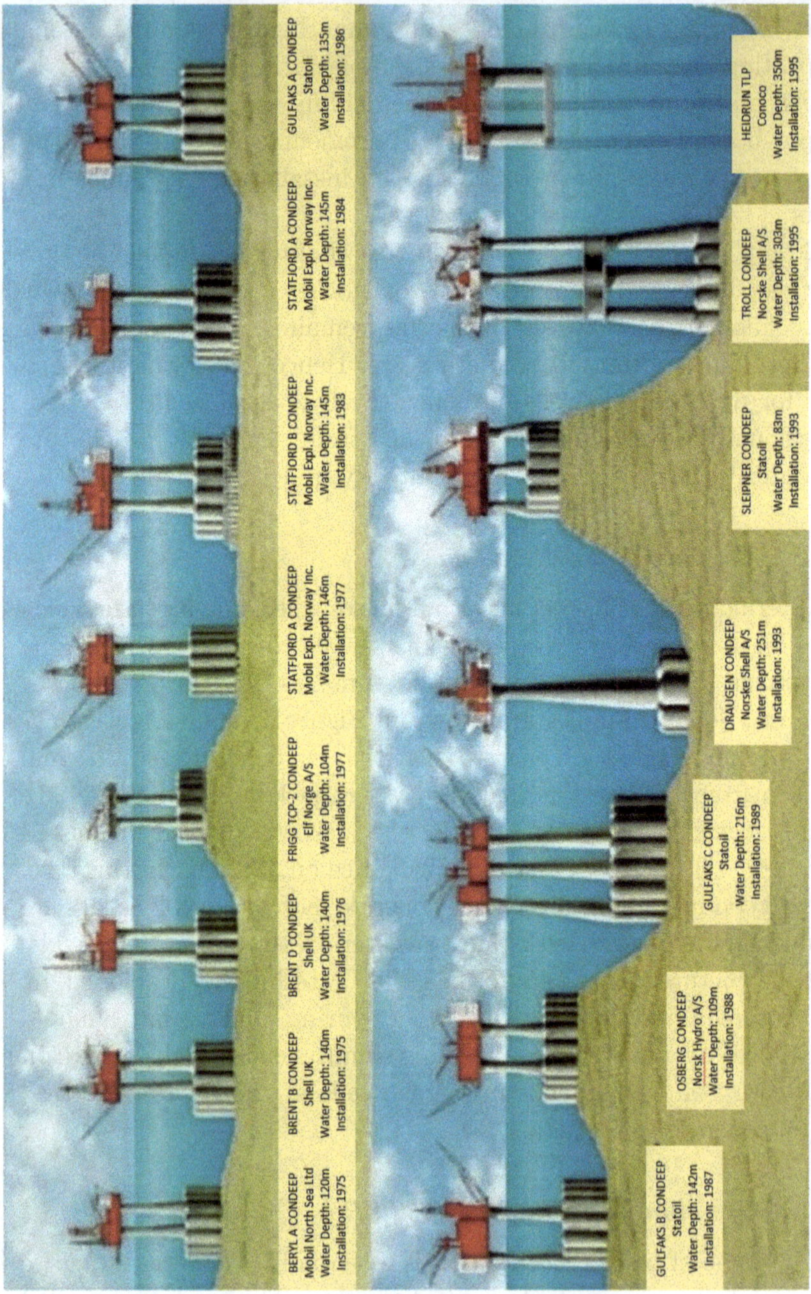

Figure 1.16. Examples of concrete jacket installations.

Figure 1.17. Concrete platform installation.

into the seabed and effectively "pull" the floating structure down against its buoyancy. The result is a stable platform that can be deployed in water depths up to 7,000 ft. A **Spar** is a cylinder-shaped vessel tied to the seabed in a similar way to the TLP.

1.6.4. *Tender-Assisted Drilling Rigs*

Once the drilling activities are complete on a platform, the rig is often mothballed until it is required to workover a well or carry out some other well operation. This is inefficient because re-activating a rig that has not been used for some time is very costly. A better solution in less harsh environments is to use a tender rig to carry out the well activities on a platform, as this can be moved elsewhere for similar operations (see Figure 1.19).

The tender is typically a moored barge or semisubmersible unit with a large crane that lifts the rig in parts or as a whole onto the

Figure 1.18. Concrete platform installation.

platform. Once on the platform, work is carried out as if the rig were permanently installed. Fluids, supplies and control of the well are provided by the tender unit. When operations are completed the rig is lifted off the jacket onto the tender and sailed to another location. This is only possible in benign wind and sea conditions due to the heavy lifting operations required.

Jackup Drilling Rigs. This is an alternative to tender operations over platforms for water depths up to 150 m, the jackup is a mobile drilling unit with (typically 3 or 4) extendable legs that sit on the

Figure 1.19. Tender assisted rig.

seabed when the rig is operating. The hull of the rig is watertight and the unit is towed onto location using tugs. Once there, the rig is positioned as close as possible to the jacket (avoiding pipelines on the seabed), the legs jacked-down onto the seabed and the hull raised above the water level. When above the top deck of the platform, the drilling derrick is skidded outboard of the rig so that it is cantilevered vertically over the well to be accessed. Again, drilling proceeds in a similar way to the drilling of any well through the platform. Single wells, such as exploration wells, can also be drilled without the need for a jacket at all — in such conditions a conductor pipe is used and it is suspended from the jackup itself rather than by the jacket (see Figures 1.20 and 1.21).

The condition of the seabed under the legs of the Jackup is a key consideration. Surveys are carried out prior to positioning the rig to ensure that seabed is strong enough to support the rig. A further consideration at a platform location is "footprints" created by previous jack-up rigs with different leg layouts and sizes. Once jacked-up, a load test is carried out by using water ballast on the rig

Figure 1.20. Jackup rig working over steel jacket.

Figure 1.21. Jackup rig working over single well.

to ensure seabed integrity. There have been numerous examples in the past when jackup rigs have been damaged due to weak seabed conditions.

Jackup and semisubmersible rigs are moved short distances and positioned using tugs. For longer distances, it is more normal to use semisubmersible transport vessels (see Figures 1.22 and 1.23). Once the vessel is ballasted down, the rig is floated over it, whereupon the vessel can be ballasted-up to operating draft and sailed carrying the jackup to the new location.

1.6.5. *Semisubmersibles*

Semisubmersible drilling units entered service in the early 1960s but have been developed intensively since then. These rigs are able to sail between locations at a shallow draft and once on location are ballasted down to improve their stability in rough seas. The rigs are held in position either using anchors (typically 8 or 12) laid out in a "star" pattern or Dynamic Positioning (DP).

Figure 1.22. Jackup transportation by semisubmersible ship.

Figure 1.23. Rig transportation by semisubmersible ship.

The topsides of the rig are very similar to that of a jackup, with three important exceptions:

- The drilling derrick needs to compensate for vertical motion of the rig relative to the seabed.
- The BOPs for the semisubmersible rig are located on the seabed and hence are deployed, operated and maintained in a different manner to those on the jackup.
- The wellhead for a subsea well (i.e. one drilled by a semisubmersible) is at the seabed rather than at surface.

Newer semisubmersible, jackup and drillship rigs tend to be equipped with dual-activity derricks (see Figures 1.24 and 1.25) and additional equipment one deck below that can shuttle pipe and BOPs between derrick centres. This allows simultaneous operations to take place, such as running marine riser while making-up a drilling assembly. Each derrick has its own crew.

Both semisubmersibles and drillships have "**moon-pools**" that are areas open to the sea under the drill floor(s) for the running of

Figure 1.24. Sixth-generation semi-submersible drilling unit.

large equipment such as BOPs and subsea Christmas trees. This also houses the system needed to maintain the marine riser under tension, while accommodating vessel movement, heave, yaw and roll.

Like other rig types, building of new semisubmersibles has tended to take place in "waves" that corresponded to periods of high oil prices and market optimism by drilling contractors. Consequently, semisubmersible rigs are sometimes described as fitting into one of several generations are shown in this table:

Generation	Approximate water depth (ft/m)	Dates	Typical day-rate in 2018 \$k/d
First	600/200	Early 1960's	Not available
Second	1,000/300	1969–1974	Not available
Third	1,500/500	Early 1980's	Not available
Fourth	3,000/1,000	1990's	150
Fifth	7,500/2,500	1998–2004	200
Sixth	10,000/3,000	2005–2011	250

Figure 1.25. Dual activity derrick.

Considerations for the design of the semisubmersible drilling rig are maximum operating water depth, maximum deck loading and the topside equipment package. Some semisubmersible rigs and drillships are equipped with Dynamic Positioning (DP) systems. These comprise typically 4–8 motor/propeller azimuthing thrusters that can rotate by 360°. The propellers are fixed or of variable pitch, and the units can be replaced as a single component (see Figure 1.26). These devices enable the rig to be positioned vertically over the well at all times, overcoming current and wind forces. They are controlled by a sophisticated computer system that takes input from GPS, beacons on the seabed and other sensors. High reliability of these systems is essential for the successful deployment of this type of rig. DP systems also allow the rig to be "weather-vaned" into the

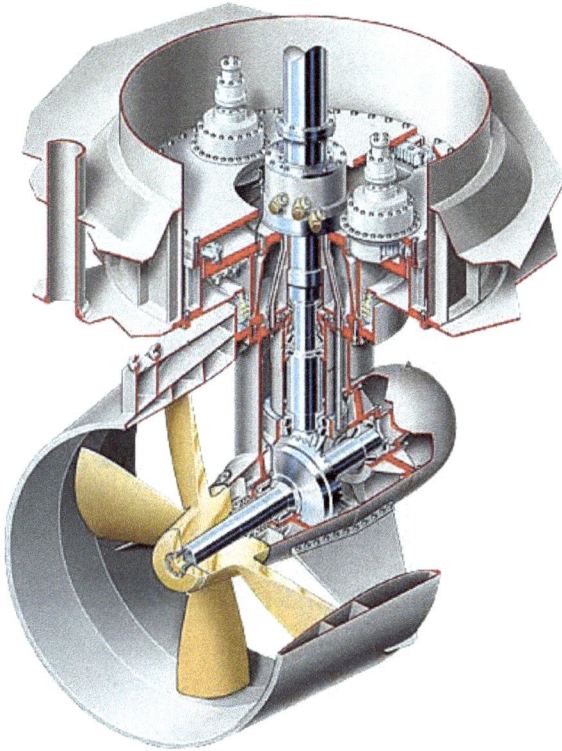

Figure 1.26. Azimuthing thruster.

sea or wind conditions, to minimise heave and roll. Drillships (see Section 1.6.6) and other oilfield vessels such as large supply boats, Floating Production, Storage and Offloading Systems (FPSOs), Diving Support Vessels (DSVs) and heavy-lift vessels use similar technology.

In the event of very poor weather, a severe well control emergency or risk of collision with another vessel, the semisubmersible rig (and drillship) is designed to disconnect from the well in a safe manner.

1.6.6. *Drillship*

Drillships have a number of advantages over semisubmersibles. They can move more quickly between locations and have a very significant deck-load capacity. This translates into an ability to drill

Figure 1.27. Dual activity drillship.

an entire well, or even a series of wells, without replenishment of supplies. The downside is that they are more weather-sensitive than a semisubmersible, due to their hull design (see Figure 1.27).

1.6.7. *Coiled Tubing Rig*

As will become clear later in this chapter, conventional drilling involves turning drill pipe at surface. All rigs described thus far support this operation. Over the last 20–30 years, another type of drilling method based on non-rotating continuous steel tube (a technique called Coiled Tubing Drilling or CTD) has been introduced. CTD rigs such as the one pictured in Figure 1.28 are typically truck-mounted.

The main components are:

1. a reel of coiled tubing, typically $1\frac{1}{2}''$–$3''$ in diameter and up to 15,000 ft in length;

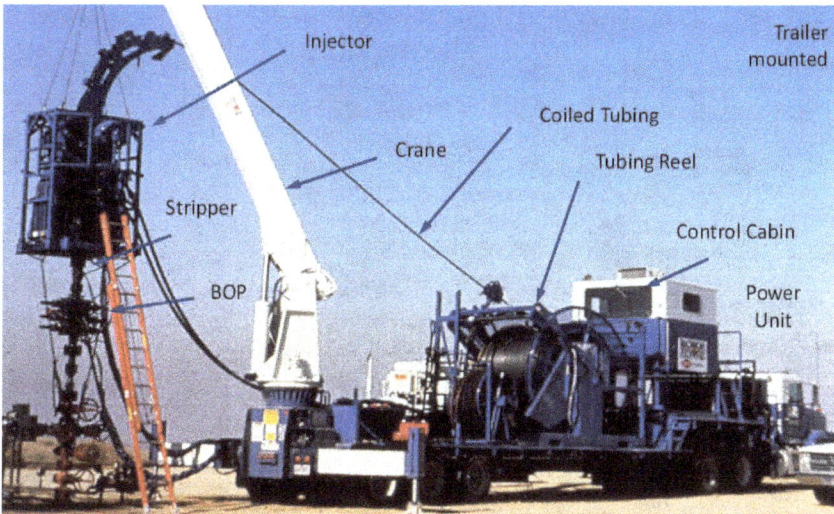

Figure 1.28. Coiled tubing rig.

2. an injector that can push tubing into the well against pressure or friction resistance;
3. a stripper that seals the space between the coiled tubing and wellhead;
4. a BlowOut Preventer (BOP) that allows the well to be sealed in an emergency;
5. a crane to support items 2–4 above;
6. a power unit for running the coiled tubing into or out of the well;
7. a control cabin for the operators to manage the power and control functions.

These individual components will be covered later in this chapter.

Coiled tubing rigs are also deployed on offshore platforms and for non-drilling activities such as running completions, stimulating, perforating and servicing a well. They are of lower cost and have fewer crew than full drilling units, and recent developments mean that their capabilities are increasing.

Figure 1.29 shows a coiled tubing unit rigged up on a conventional rig floor.

Figure 1.29. Coiled tubing rigged-up on land rig.

1.6.8.   *Hydraulic Snubbing Unit*

A snubbing unit is a rig-like device used to run pipe into or out of a well that is already under pressure. This may be required to workover a well that cannot be killed perhaps due to tubular failure which precludes circulating kill-weight fluid. The situation is complicated by the need to push tubulars into a well — their weight will not be sufficient to overcome any significant well pressure. Hydraulic rams are typically used to create this force, at least for the first few tubulars until sufficient weight in the hole has been achieved. To seal around the tubulars, a series of sealing rams and BOPs are used whose operation must synchronise with the hydraulic rams. The total "stack-up" height can be quite considerable (see Figure 1.30).

1.7.   **Function of the Rig**

All drilling rigs have the following four basic functions:

1. hoist & lower — tubulars into and out of the well;
2. rotate — the drilling string to drill;

Figure 1.30. Hydraulic snubbing unit.

3. circulate — pump fluids down the drill string and up the well;
4. control pressure — prevent unwanted flow from the well.

Figure 1.31 illustrates the most basic of drilling rigs. The rig is positioned vertically above the location where the hole is to be drilled. Drilling involves rotating a drill bit in the hole on a drill string that comprises pipes connected back to the surface. The pipes allow circulation of drilling fluid to flush out the rock pieces drilled by the bit.

1.7.1. *Hoist and Lower*

To lower the bit into the well, a derrick is required — which is usually a vertical tower comprising steel beams bolted together, positioned above a **drill floor** or **rig floor**. This also allows the bit to be pulled back to surface to be replaced when it is worn.

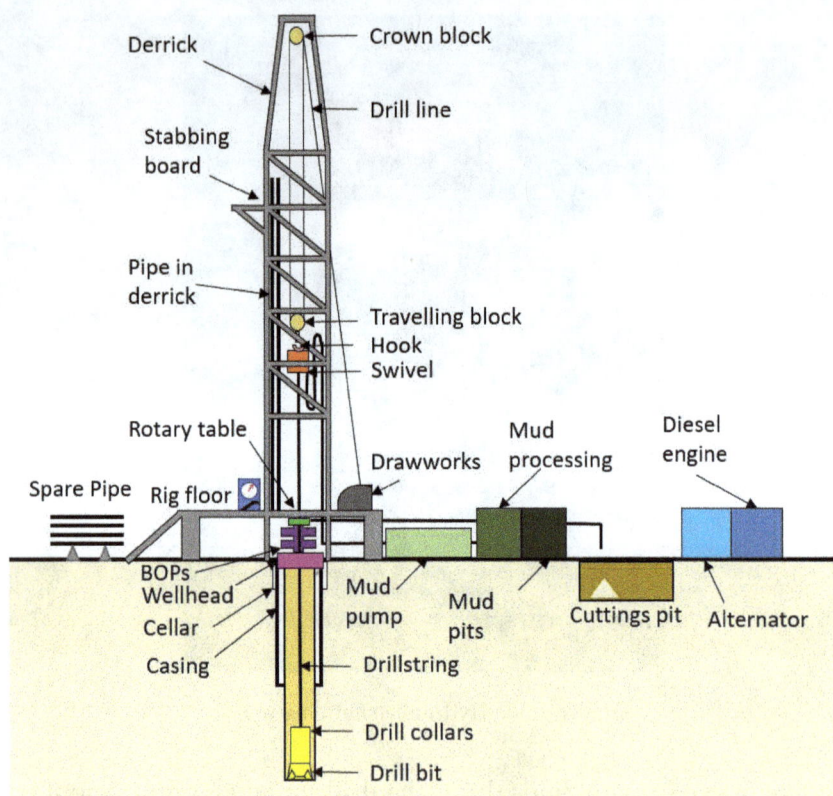

Figure 1.31. Basic drilling rig.

This pulling and rerunning of the bit on the end of the drill string is called *tripping*. When tripping out the drill string, pipes are unscrewed and kept vertical in the derrick — this process is called *racking-back*.

The reverse happens when running back into the hole. In the basic drilling rig, a block-and-tackle arrangement is used to raise and lower the drill string — the **crown block** is attached to the top of the derrick structure and the **travelling block** moves vertically within the derrick structure. Attached to the travelling block is a large hook that can be connected to and disconnected from the drill string. Both crown and travelling blocks comprise typically 4–8 pulley wheels that are strung with steel cable known as the **drill line**. The drill line is connected at one end to a wide pulley (or **drum**) on the draw-works, which reels it in and out as required to raise and lower the travelling block. The other end of the drill line is anchored to the drill floor by a clamp called the **dead line anchor**. This arrangement provides gearing so that tension on the draw-works is less than the weight of the drill string, reduces the size of the drill line and provides finer control over pipe movement.

The draw-works is powered by one or more electric motors, equipped with one or more braking systems and a system to spool the drill line correctly onto the drum. This assembly is attached to the drill floor. The motor is used to raise the drill string and the brake is used to control the speed of lowering it. The brake is also an important component to control the drilling process by adjusting the vertical load on the bit — called the **Weight on Bit** (WOB).

The load on the drill string is measured by a sensor on the dead line anchor. There are various sensors in the derrick to measure the position of the travelling block and to prevent it colliding with the crown or the drill floor. The drill line needs to be replaced due to wear on a regular basis determined by hook load and travel — this is done by slipping line through the dead line anchor. Derrick equipment is shown in Figures 1.32(a)–1.32(e) and the hook arrangement for a rig without a top drive is shown in Figure 1.33.

Derricks are fixed on large rigs, but on smaller units can be extended and retracted when moving the rig. All lifting devices need to be regularly inspected and checked to ensure there are no loose items that could fall and injure someone on the rig floor. In many cases (e.g. on semisubmersibles and drillships), the travelling block runs on vertical tracks in the derrick structure to prevent uncontrolled lateral movement that might arise, e.g. from rig motion

Figure 1.32. Derrick lifting equipment clockwise from top left: (a) travelling block and hook, (b) crown block, (c) drawworks, (d) dead line anchor and (e) driller's position.

on the sea. Where a top drive is installed, this is also needed to transfer drill string torque to the derrick.

Derricks need to withstand the vertical loads of the drill string or other equipment being used in the well, side loads arising from wind and in some cases torsional loads from rotating the drill string. Facilities are provided inside the derrick for a crew member (the *Derrickman*) to screw and unscrew the drill string, manage the racking-back of pipe and maintain equipment. The pipe is normally racked-back in **stands** of 2–4 drill pipes that remain screwed together. In cold climates, derricks are enclosed to maintain a suitable working environment.

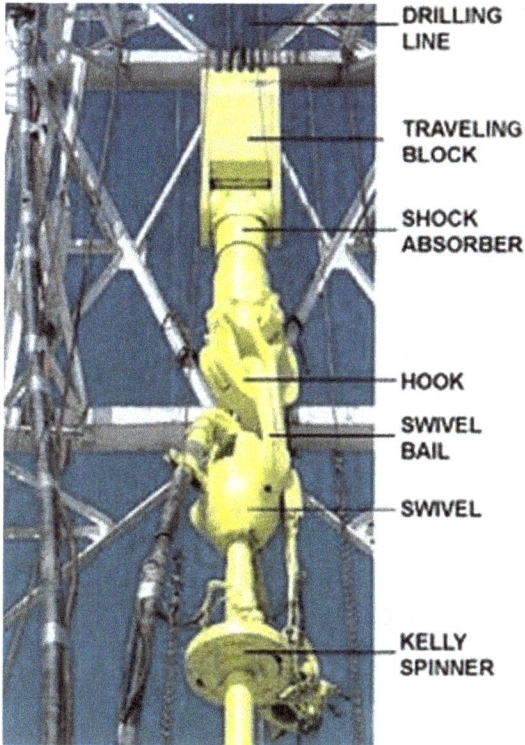

Figure 1.33. Hook arrangement (no top drive).

Lighter rigs use a mast rather than a derrick. In plan view, a derrick surrounds the drill centre, whereas a mast is erected to one side. The mast is generally lighter and designed to be more easily extendable; however, it needs to accommodate the offset load resulting in bending moment on the structure.

1.7.2. *Rotate*

An essential requirement for all drilling rigs is that the bit be rotated to cut rock from the bottom of the well. Bit rotational speed (measured in Revolutions Per Minute or RPM) and Weight On Bit (WOB) are key parameters in determining the efficiency of drilling. In most drilling, the entire drill string needs to be rotated from surface to turn the bit.

Historically the drill string has been rotated by a *turntable* that fits into the drill floor and is driven by an engine or more usually by an electric motor. The turntable rotates a *kelly bushing* which in turn rotates a tubular in the drill string called a kelly that is hexagonal in cross-section. This arrangement allows the string to be rotated while simultaneously allowing vertical movement of the drill string. This allows the bit to drill down the hole while rotating (see Figures 1.34 and 1.35).

The rotary table can be set for clockwise or anticlockwise rotation and adjusted for speed, and is equipped with sensors to measure RPM and the torque being applied to the drill string. Drilling itself always takes place in a clockwise direction (when viewed from above) — often called **turning to the right**.

One significant development in the last 20 years has been the near-universal application of **top-drive systems** on all but the simplest drilling units. The top-drive replaces the rotary table with a drive system suspended between the hook and the top of the drill string. Top drives (see Figure 1.36) can be either hydraulically or (more usually) electrically driven. They rotate the drill string directly at any time (for example when tripping and running casing), and allow drilling without interruption of a stand (2, 3 or 4 drill pipes).

1.7.3.   *Circulate*

As we have seen, a basic requirement of the drilling rig is the circulation of drilling fluid in the well. Drilling fluids will be covered later in this chapter. For the time being, note the following basic requirements of the drilling fluid:

- remove drilled cuttings;
- add drilling power at the bit;
- clean the bottom of the hole;
- cool the bit.

For most operations, the drilling fluid circulates around a closed-system so that essentially it is re-used continuously.

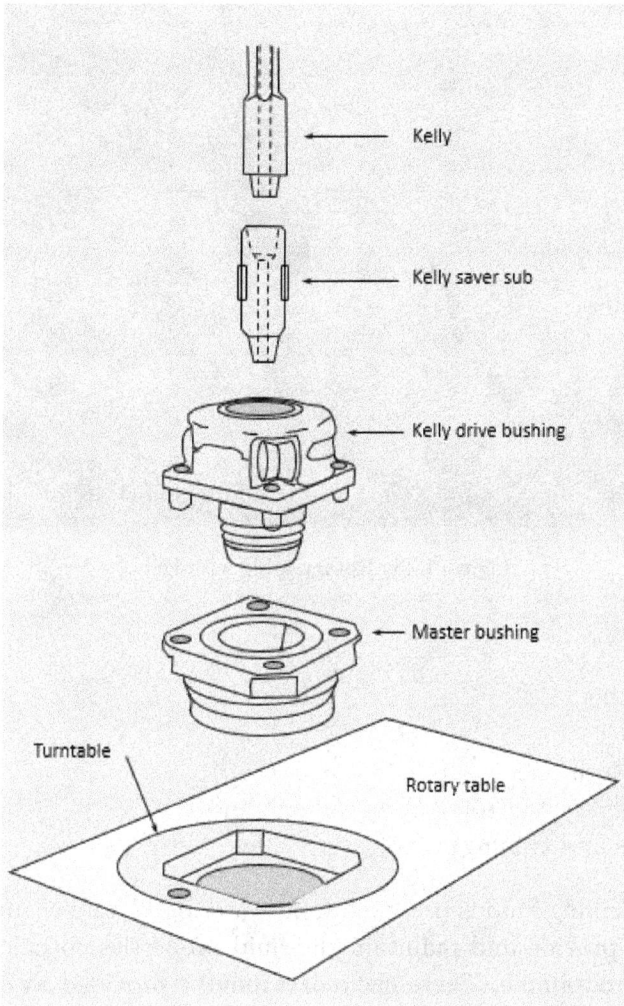

Figure 1.34. Rotary table arrangement.

The circulation system comprises the following that will be covered in this section (see also Figure 1.37):

- mud pits and mixing systems;
- mud pump;
- standpipe/rotary hose;

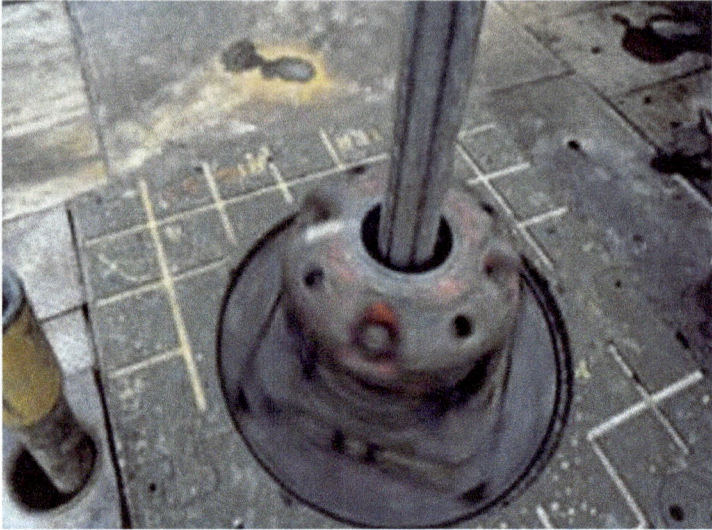

Figure 1.35. Rotary table with kelly.

- swivel;
- drill string;
- borehole;
- mud return line;
- shale shaker and solids removal equipment;
- reserves and cuttings pits.

The drilling fluid is prepared from water or oil, and chemicals are added to provide and maintain the fluid properties specified in the drilling programme. These chemicals may be provided as a powder in bulk tanks or trucks, or in sacks, or in drums containing liquids. Typically, a hopper is used to add the chemicals, and pumps are used to treat the mud offline from the main circulation path.

The mud pits comprise a series of metal tanks linked together with valves or weirs between them, and with pumps to allow circulation. Usually, each tank is equipped with a stirrer driven by an electric motor, and each has one or more level sensors to determine the volume of liquid at any time. A grating covers the top

Figure 1.36. Top-drive system.

of each tank to allow visual inspection and samples to be taken for testing. Some pits contain drilling fluid in the circulation system (the *active system*), while others contain fluids in reserve. Mud is drawn from the base of the active system and pumped to the main mud pump.

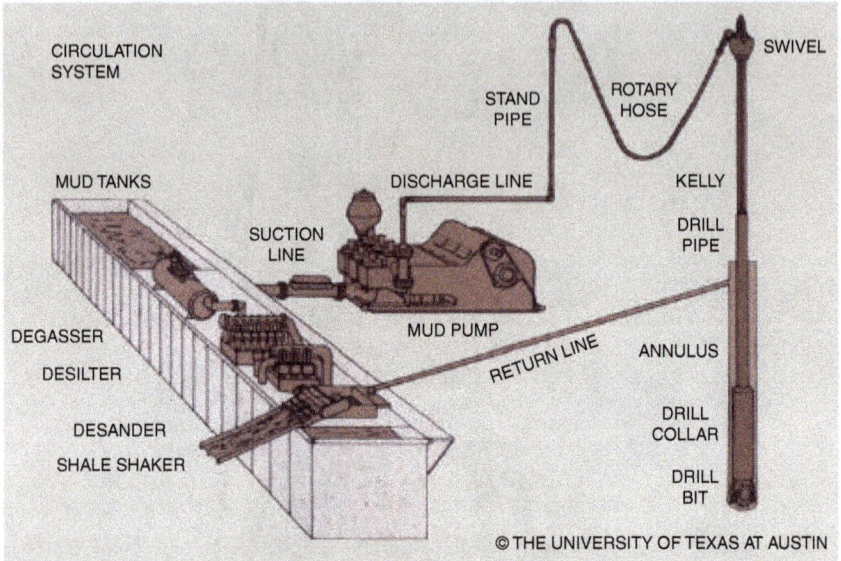

Figure 1.37. Circulation system.

The mud pump provides the increase in fluid pressure required to circulate the liquid around the well. Typically, mud pumps are a reciprocating design with three cylinders, called a *triplex* pump (see Figure 1.38). This allows pressures of 5,000–7,500 psi to be attained while reducing pressure surges in the system. These surges are further reduced by using a *pulsation dampener* — a type of air-filled bladder — on the outlet and inlet sides of the pump. Pumps have the following features:

- can handle fluids containing high percentages of (abrasive) solids;
- allow passage of large solid particles (typically lost circulation materials);
- simple operation and maintenance. Liners, pistons and valves can be replaced in the field by the rig crew;
- capable of pumping a wide range of flow rates and pressures using different liner and piston sizes.

Typically, 2–4 pumps are available to provide redundancy so that reconfigurations and repairs can take place while circulating.

Figure 1.38. Triplex pump.

From the mud pump, the drilling fluids pass via a **stand pipe manifold** up the derrick in a **stand pipe**, and through a high-pressure flexible hose into the drill string via a *swivel*. The swivel allows the fluid to enter the drill sting while it is suspended in the derrick and while it is rotating.

Thereafter, the drilling fluid flows down the inside of the drill string and through the bit. Here it loses pressure and picks up the rock cuttings that have been produced by the drilling process. Flow returns to the drilling rig up the annulus, between the hole that has been cut and the drill string. Once back at the rig, now at atmospheric pressure, the drilling fluid returns to the active pits via a mud return flowline and shale shakers which are vibrating screens designed to remove the drilled cuttings.

The drilled cuttings so removed are channelled into a **cuttings pit** that on land is typically a hole in the ground alongside the drilling rig that has been lined to prevent ground contamination. On completion of the well, cuttings are either buried *in situ* or more often removed and disposed of in a controlled process offsite.

There are a multitude of other devices connected to the circulation system to treat the fluids.

The **shale shaker** comprises a series of steel filter screens of selectable mesh that vibrate under a cascade of drilling fluid to remove the majority of drilled cuttings. Any entrained gas and fumes from the mud are removed from the working area around the shakers by extraction systems on modern rigs. The well-site geologist takes the cuttings samples for analysis from the shakers.

The **desander** and **desilter** are devices that are fed by pumps which rely on centrifugal effects of flow inside a cone to remove sand and silt from the drilling fluid. A **centrifuge** comprises a rotating drum that removes the lowest density solids from the fluid. The **degasser** is a vacuum vessel in which a partial vacuum is used to extract gas that tends to remain in solution in the drilling fluid. These latter devices are not used in the main flow path but work on the fluid while in the active system of pits.

The system is flexible to allow, for example, *reverse circulation* — that is down the annulus and up the drill pipe if that should be required.

Once again, pressures and flow-rates are measured by a multitude of sensors around the system. An important consideration when running the drill string into and out of the well is that the well be kept full of drilling fluid, but not allowed to flow. A tall narrow tank called a *trip tank* is used to monitor these fluid changes, and an important duty for the driller is to ensure that changes in this tank level exactly correspond to the volume of pipe being removed from or run into the well.

1.7.4. *Control*

The fourth and last main component of the drilling rig is control. By this we mean taking measures to ensure that the well is not allowed to flow in an uncontrolled way. Such an event would (and has) led to serious injury, loss of equipment and reputation and severe environmental damage.

A well is a pressure containment system. If the pressure in the well gets too great, there is a danger that leaks will take place. The well and rig are designed to detect and withstand maximum expected pressures with a safety factor. Barriers are put in place to safeguard the well, and during operations there must be a least two barriers in place at any time. Barriers include drilling fluids of sufficient density, BOPs, valves, casing and cement.

The simplest well control element is called the **diverter system**, and this is used in the shallow sections of the well — i.e. at the start of drilling operations. It is not a barrier but is intended to divert the flow of the drilling fluid and any hydrocarbon liquids or gases away from the drilling rig in a controlled way to allow people on the drilling rig to leave the area safely (see Figure 1.39).

The diverter comprises a rubber element that closes in the annulus between the hole and drill string, while simultaneously opening a value to allow the fluids to flow from under the rig floor along a pipe to the cuttings pit or overboard if offshore. The diverter pipe needs to be of large diameter to allow rapid flow of fluid and formation of solids with minimal pressure drop.

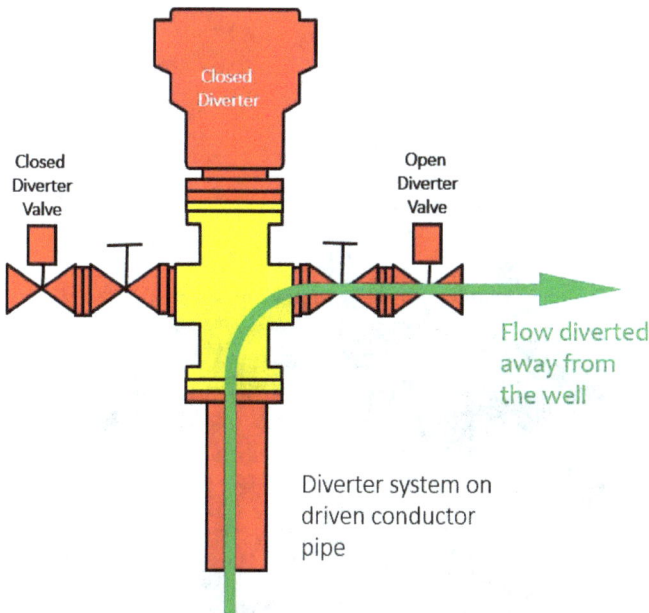

Figure 1.39. Diverter stack-up.

BOPs are essentially a series of valves designed to close-in the well if unexpected flow occurs in the well. There are two basic versions — **the ram-type** preventer and the **annular** preventer. The latter is rather like the diverter already described in that it comprises a rubber element that is moved into the annulus space around the drill string. These are typically rated for 2,000–10,000 psi and have the advantage that they can seal against a wide variety of drill string diameters. They also allow — under controlled conditions — **stripping** of tubulars through them with pressure below the preventer. This is used for certain well control operations (see Figure 1.40).

Figure 1.40. Annular preventer.

Figure 1.41. Pipe ram section.

Ram-type BOPs (see Figure 1.41) comprise a pair of rams that are hydraulically actuated to close around the drill string and retain pressure from below — ratings are typically 5,000–20,000 psi. These are usually sized for specific pipe diameters, although some (variable bore rams) can accommodate a range of diameters (a smaller range than an annular). Shear rams are designed to shear the drill string in an emergency, and blind/shear rams are designed to shear pipe and seal off the entire wellbore cross-section thereafter. Typically, a combination of ram and annular BOPs are combined in a single package — called the BOP *stack*. Ram preventers are

Figure 1.42. BOP hydraulic (Koomey) control unit.

designed to be tolerant of pipe movement and above all need to be reliable.

Controls for BOPs are hydraulic or electro-hydraulic and can usually be operated from several locations such as the drill floor and wellsite office. An **accumulator system** (see Figure 1.42) allows operation based on compressed gas even in the event of an electrical power failure. Land, jackup or platform based operations typically use an annular and three rams. Between these devices access pipes and valves are connected (called **choke and kill lines**) to allow circulation in the well (see Figure 1.43).

For semisubmersible and drillship applications, a more sophisticated BOP stack is required, as this will typically sit on the wellhead at the seabed. This BOP is run on a riser from the drilling rig and includes a disconnection facility so that the rig can depart the location in an emergency. The BOP is also remotely connected to the wellhead. Due to the water depths involved, more complex control systems are required to ensure rapid and reliable operation. In these applications, 2 annular and 4–6 ram BOPs are used, and additional accumulator capacity is located at the seabed. In some cases, they can be operated using a **Remote Operating Vehicle (ROV)**.

Flow up inside the drill string is less common but is prevented using non-return valves in the pipe and by using a valve in the string at surface called an **Internal BOP (IBOP)**.

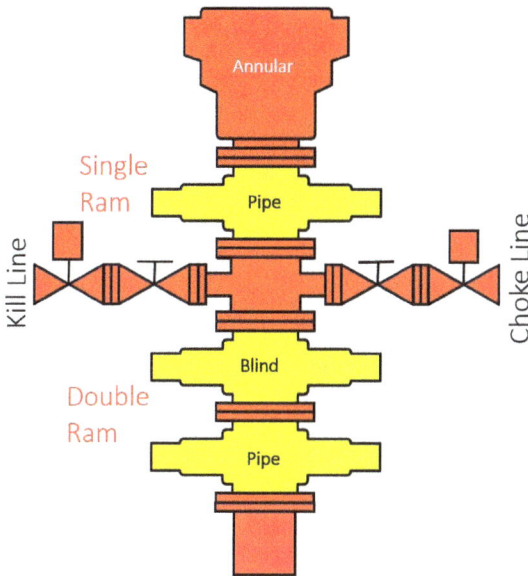

Figure 1.43. Pipe ram stack-up.

The decision to close-in the well in an emergency is made by the Driller, and must be done without delay to minimise the size of any fluid influx. He will determine flow from the well visually or by using a variety of instruments that determine flow and/or an increase in the total drilling fluid volume at surface. Typically, a ram-type preventer will be closed and the drill string landed off so that its weight rests on the ram.

The process for circulating out a kick is well defined and most of the calculations will already have been made by the driller and his supervisors on the wellsite (and noted on a **kick sheet**). Killing a well comprises circulating the influx out by controlling the pressure on the annulus side of the well to maintain a constant bottom-hole pressure that is greater than the pore pressure. Instead of taking returns to the shale shaker and mud pits, the return flow is channelled through a manifold and a choke, thereafter to a **mud–gas separator** (see Figure 1.44) to allow any gas in the mud to be liberated before it is returned to the pits. The choke is managed remotely from a panel on the drill floor and comprises a valve designed to regulate the flow of fluid.

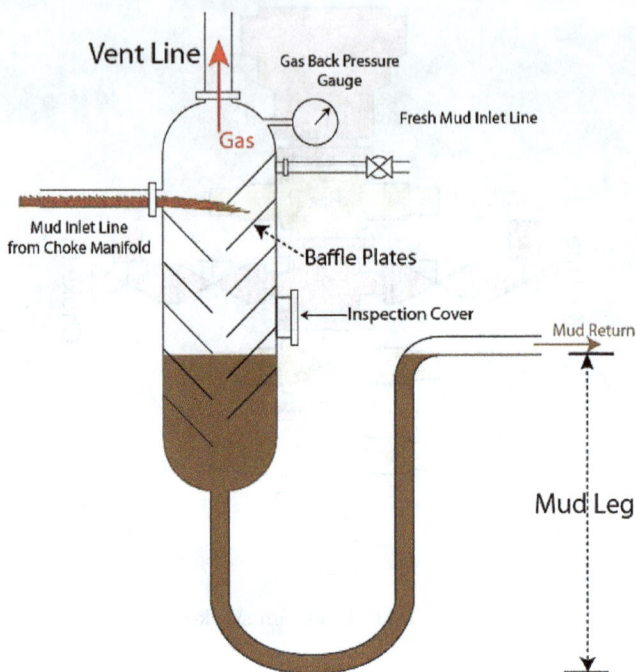

Figure 1.44. Mud gas separator.

1.8. Drill Floor Equipment

A typical drilling rig floor includes numerous pieces of specialist equipment to support operations. Early rig development took place in North America, and as a result many of the terms used to describe operations, items of equipment and crew roles are based on American life — such as mouse-hole, dog-house, cat-head and red-neck. In this section, we will describe some typical features of the rig floor.

The Driller is central to drill floor operations. He is responsible for safe operations of the rig and his crew. He is also accountable for what is run into the well — lengths, ODs, IDs and profiles of all BHA components. The ID of all pipes is checked with a drift that is dropped down inside each stand in the derrick.

Slips are a set of wedges that are positioned around the drill string to support it in the rotary table on the drill floor, and are sized to fit the pipe's outside diameter. They are typically installed

Figure 1.45. Setting slips manually.

and removed by members of the drilling crew on the instructions of the driller (see Figure 1.45). **Elevators** are used to support the top of the drill string from the hook. There are several different support methods — slips (as above) or a load-bearing shoulder. The elevators are supported by the hook using *bales* (short supporting rods) which are connected to lugs. Elevators are built in two halves hinged together to allow them to be opened and closed around the drill string. A type of elevator for running casing is shown in Figure 1.46.

Screwing and unscrewing drill string tubulars (called *making-up* and *breaking-out a connection*) requires a number of pieces of equipment to hold the pipe and to apply the high torques required. A **chain tong** is the simplest manual device but is limited to torque that can be applied by 2 or 3 men. The **spinning chain** was used for years to quickly make connections but was notoriously dangerous. The basic method used today is the **make-up/break-out tong** that grips the pipe body and is pulled sideways by a cable from

Figure 1.46. Automated casing running tools.

the draw-works cat-head; the torque is opposed by a second tong connected to a fixed post on the drill floor, with a load sensor calibrated to indicate the torque applied. Modern rigs are equipped with an **iron roughneck**, which is a device containing two wrenches that are positioned either side of the connection and which apply the required torque under remote control — see Figure 1.47. This machine can be moved up to and away from the drill string on tracks as required. The focus on these and other operations is to automate where possible to reduce human exposure to physical risks.

As has been mentioned already, drilling tubulars are frequently stored (called *racking-back*) in the derrick, for example to pull the drilling bit out of the well for replacement (see Figure 1.48). Historically, the drill string has been unscrewed in sections, moved to the side of the derrick with the upper end of the pipe secured

Figure 1.47. Iron roughneck.

in derrick *fingerboards* for support. This operation has been done manually, but requires a skilled person (the Derrickman) to be located some 100 ft up in the derrick. Newer rig designs have automatic racking systems to manage these activities and keep track of the pipe inventory.

When a new pipe needs to be brought into the derrick, automated systems are replacing manual pipe handling. Prior to loading tubulars into the derrick, they are organised, measured and inspected on a horizontal area called the pipe deck. On a land rig, the racking area is adjacent to an area where trucks can deliver the pipe — it is usually offloaded by crane or fork-lift truck.

For general purpose lifting on the rig-floor, a set of winches (**tuggers**) are used — these are powered by compressed air.

Older and more modern types of drillers cabins are shown in Figures 1.49 and 1.50, respectively.

Figure 1.48. Racking-back pipe.

Because of the ever-present risk of hydrocarbons on the rig floor and nearby areas, all electrical equipment deployed are intrinsically safe — i.e. designed not to create a spark that could cause ignition. Use of everyday electrical and electronic equipment is not allowed, and most electrical items are housed in enclosures, continuously purged by air at elevated pressures.

1.9. Drill String Components

The drill string has the following purposes:

- running and recovering the bit and other components into and out of the well;

Figure 1.49. Old–style Driller's controls.

- adjusting the weight on bit;
- forming a conduit for drilling mud;
- providing stiffness to ensure that the bit drills in the required direction.

The use of high-strength steels in the drill string supported significant developments in drilling operations in the period 1940–1970.

The basic drill string elements are shown in Figure 1.51 — in this case for Kelly drilling.

We have already covered most of the drill string items above the rotary table. The **swivel** supports the entire drill string off the hook, while allowing it to rotate. The **Kelly cock** contains a valve

Figure 1.50. New style Driller's controls.

that allows the drill string to be isolated, for example, in the case of uncontrolled flow up the drill string.

Below the drill floor, the drill string comprises mostly **drill pipe** (covered in more detail below). Just above the bit, as part of what is called the Bottom Hole Assembly (BHA), several **drill collars** are also used. Drill collars are thick-walled steel tubulars that provide the weight on bit that is required for drilling, and they are designed to be rotated in compression (as opposed to the drill pipe itself that is usually in tension). See the tension load chart on the right side of Figure 1.51. Throughout the drill string, short pipe sections called **subs** are used between the major components so that key threaded connections can remain intact when running and recovering the drill string. A **crossover** tubular is used between different diameters of tubular or different types of thread.

Between each tubular member, there is a threaded connection that must be able to withstand tension, bending and pressure loads,

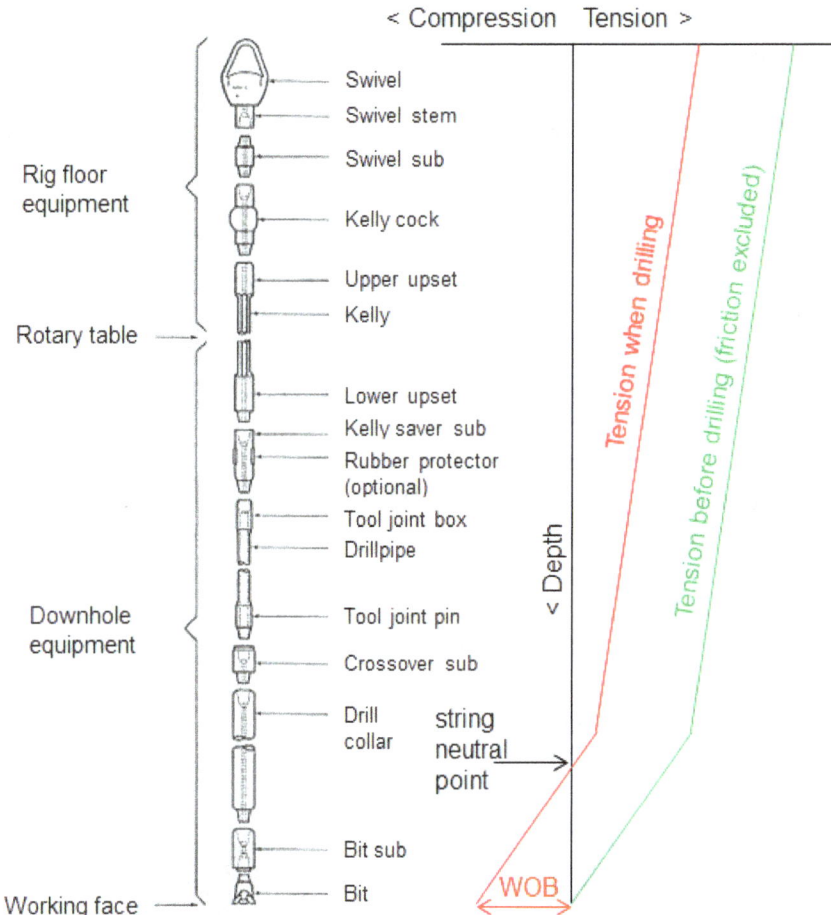

Figure 1.51. Drill string components and tension.

while sealing drilling fluid being pumped through it. These threaded connections comprise a male (the **pin**) and female (the **box**) sides and rely on metal-to-metal seal faces rather than elastomers for sealing. Make-up torque for each connection is specified and closely managed by the Driller, and grease (called **pipe dope**) is manually applied when making the connection for lubrication.

Tubulars come in a wide range of lengths and Outside/Inside Diameters (**OD**s/**ID**s). A single drill pipe is typically 30 ft or 40 ft in length, and between $3\frac{1}{2}''$ and $6\frac{5}{8}''$ OD. A common size is 5''

OD, 4.276″ ID, which equates to a linear weight of 19.5 lb/ft. Drill pipe can be supplied in various grades (e.g. E, X, G, S) that reflect different yield strengths of the steel. The pipe is made in the factory by welding **tool-joints** on either end of the basic steel tube; the connections are machined in the tool joints. The tool joints have a slightly larger OD than the pipe body (called an **External Upset, EU**) and provide a surface for the application of drill floor tongs to apply the make-up and break-out torque. The surface of the tool joint is usually smoothly coated with hard-wearing material such as tungsten carbide to prevent excessive wear when in the wellbore. Drill pipe specifications are laid down in API standard 5 DP. See Figures 1.52 and 1.53.

Drill collars range from ODs of 11″ down to $3\frac{1}{8}$″ — a typical example would be 8″ OD, $2\frac{13}{16}$″ ID — a linear weight of 151 lb/ft. The combination of bending and rotation of the drill string makes fatigue failure a significant risk; the connection is designed for this, but frequent Non-Destructive Testing (NDT) and quality assurance inspections are carried out during manufacture and throughout the lifetime of the tubulars. Drill collars frequently feature spiral grooves on the outside surface — this is to reduce the risk of differential

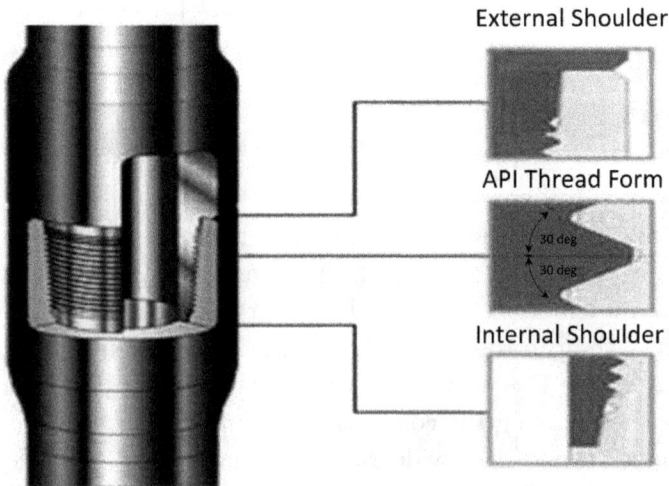

Figure 1.52. Drill pipe connections — Vallourec VAM® CDS™.

Figure 1.53. Drill pipe tool-joints.

sticking between collar and borehole wall arising from collars' large diameter — (see Figure 1.54). Measuring the direction of a well (see below) frequently requires use of one or more non-magnetic drill collars; these are made of (expensive) stainless steels.

1.9.1. *Bottom Hole Assembly (BHA)*

In reality, the BHA is usually more complex than described above — (see Figure 1.55). Typically, it additionally comprises multiple components to provide:

- downhole rotational power to the bit;
- stability;
- directional control;
- geo-steering;
- bit dynamics control;
- measurement systems.

Figure 1.54. Spiral drill collars.

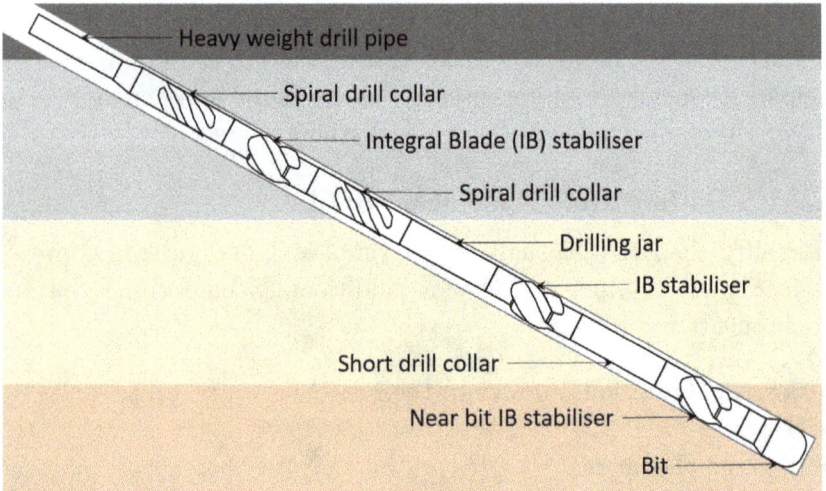

Figure 1.55. BHA components.

The following components provide this additional functionality:

The **Drilling jar** is a tool that translates changes in the upward pull on the drill string into an impulse or jarring action that is used to free the drill string in the event that is gets stuck. It comprises a hydraulic piston arrangement. Jars need to be inspected after a prescribed number of hours in the well.

String stabilisers are a set of blades placed around the BHA with an OD similar to the ID of the wellbore being drilled (see Figure 1.56). They have a spiral profile to avoid differential sticking, while providing a conduit for mud flow in the annulus between BHA and wellbore. Stabilisers are used to centralise the BHA in the wellbore, to keep the hole open and to enable the well to be drilled straight or in any particular direction. The precise diameter and position of the stabilisers along the BHA are important.

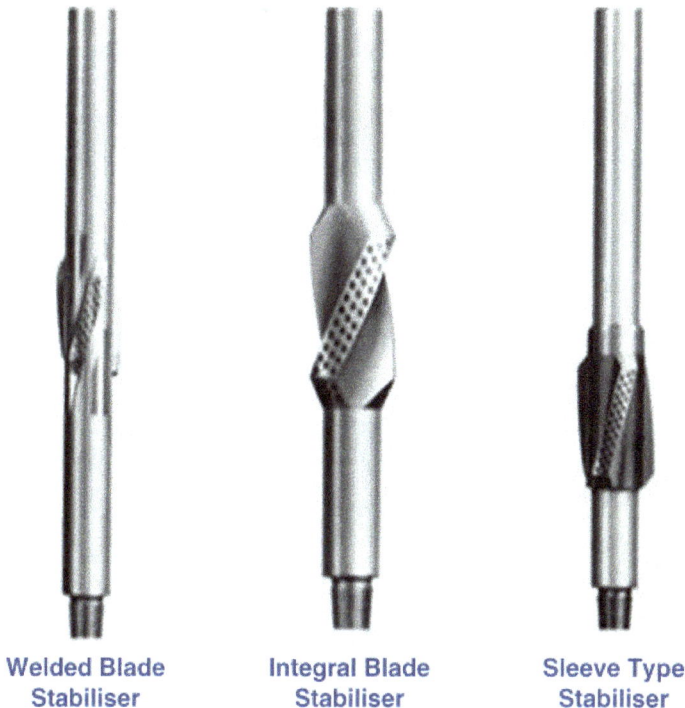

Welded Blade Stabiliser   **Integral Blade Stabiliser**   **Sleeve Type Stabiliser**

Figure 1.56. Types of stabilisers.

Typically, between one and five stabilisers are used on the BHA for drilling.

Drill string float valves are used at the bottom of the BHA just above the bit. The valve prevents reverse flow up the drill string in the event of a gas or oil influx. It comprises a flapper or plunger that is pushed and held open by the flow of drilling fluid through it. When the fluid stops or tries to go in the reverse direction, the flapper closes and holds pressure from below (see Figure 1.57).

Measurement While Drilling (MWD) tools are covered in more detail below, but suffice here to say these are stainless steel drill collars that are fitted internally with various electronic tools to measure the direction of the wellbore (azimuth and inclination) and optionally downhole drilling parameters such as bit RPM and torque, WOB, downhole vibration and pressure both inside and outside the BHA. This data is typically transmitted to surface in real-time using

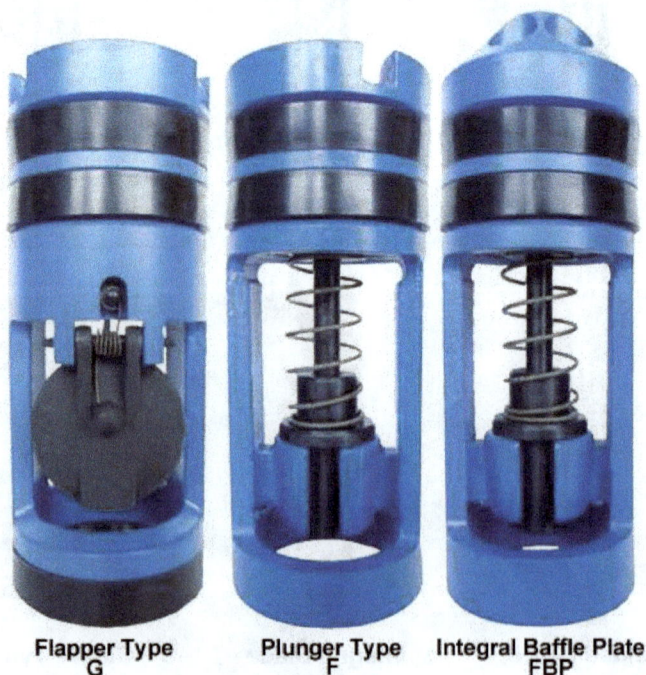

Flapper Type
G

Plunger Type
F

Integral Baffle Plate
FBP

Figure 1.57. Drill string float valves.

mud-pulse telemetry and also stored on-board the tool for later retrieval at surface. MWD directional data is used to make decisions on where to steer the drill string and to optimise the drilling process to optimise Rate Of Penetration (ROP) or wellbore stability.

Logging While Drilling (LWD) technology is similar to MWD except that it provides petrophysical data on the rock being drilled — usually Gamma Ray (GR), resistivity, porosity and density. The LWD tools are usually placed as close to the bit as possible, so that data is collected as soon as possible after drilling. Obtaining this information while actually drilling the well allows real-time decisions to be made on where to steer the wellbore, set casing, core the well etc.

Downhole motors and **turbines** are designed to turn the bit, with power developed from the flow of drilling fluid. In the case of a motor, a stainless steel rotor turns inside an elastomeric stator, as the drilling fluid flows past. A turbine relies on the fluid impacting on a series of rotors. In general, a turbine provides faster rotation (suitable for diamond bits), whereas a motor provides greater torque to the bit.

Rotary steerable systems have been developed since the early 1990s. There are a large range of technologies that "steer" or "push" the bit in a certain direction. This direction is achieved by a set of pads that push against the borehole wall or bend the drillstring in a controlled manner. The pads are actuated by hydraulic fluid that is controlled by instruction for the surface — i.e. in real time. There are also various lower cost tools that are designed, for example, to drill a totally vertical well, using MWD data and downhole computation to compensate if the well deviates from vertical.

The combination of MWD/LWD, downhole motors, rotary steerable systems (and long-life drill bits) has revolutionised drilling of wells over the last 20 years. In the most sophisticated setups, wells can be **geo-steered** so that the (notionally horizontal) wellbore remains in a specific formation. Geosteering requires surface interpretation of petrophysical and wellbore positional data against geological prognosis. A decision is made while drilling where to direct the wellbore to remain in the required formation — or in some cases a fixed distance above, e.g. the oil–water contact in a reservoir.

Very complex wellbores can now be drilled with this technology. This is further explained with examples later in this chapter.

1.9.2. *Bits*

Of all elements in the drill string, the bit is the most important. Very early drill bits relied on "percussion" drilling — i.e. repeatedly raising the bit and dropping it on the bottom of the well. The cuttings would then be dredged out of well and the process repeated.

In the early 1900s, the rotary drill bit was invented and remains in general use. The basic design is based on three "cones" that are essentially toothed wheels rotating on an inclined axis (see Figure 1.58). The axis of rotation of the cones are slightly offset so that as the bit rotates and the cones themselves rotate, the teeth shear off pieces of rock at the bottom of the hole. The pieces of rock are flushed away from the rock face into the annulus of the well by the drilling fluid. The bit contains nozzles that direct the drilling fluid onto the cutting face and the cones.

Bits are specifically designed for certain rock types. The toothed wheel described above can be machined from hardened steel or may include inserts of hardened materials such as tungsten carbide.

Figure 1.58. Tri-cone bit.

Modern bits have sealed journal or roller bearings. The bit is designed to optimise ROP and longevity. If it needs pulling after only a short distance, that is clearly not satisfactory. An important requirement is that the bit does not leave debris in the well — a common failure is that a cone comes off its bearing and is left in the well. Given the difficulty of fishing lost cones, this is best avoided! As the bit wears out (or "dulls"), the ROP slows and eventually a decision is made to pull and replace it. A heavily worn tri-cone bit is shown in Figure 1.59. If possible, the entire wellbore size section (e.g. $8^1/_2''$ hole) is drilled in a single bit-run. Bits can be reused until dull if required.

Bits are fitted with alternative nozzle sizes to allow maximisation of **Hydraulic Horse Power (HHP)** or **Jet Impact Force** at the bit to optimise drilling performance. These nozzles are fitted at surface before running the bit.

There is an industry standard system for describing bit characteristics and degree of wear.

Other common bit types are the diamond impregnated bit and the Polycrystalline Diamond Compact (PDC) bit (invented in the 1980s), as shown in Figures 1.60 and 1.61. In these types, there are

Figure 1.59. Heavily worn tri-cone bit.

Figure 1.60. Diamond bit.

no moving parts. The composite fixed cutting elements are made from artificial or real diamonds set into a body of steel. The geometry of the bit body includes flutes to channel drilling fluid and cuttings away from the rock face. The cutters themselves scrape the bottom of the well to remove rock.

Decisions on what bit to run and what drilling parameters to use are based on the following selection criteria:

- bit cost;
- rig cost/capability;
- formation;
- information requirements (e.g. are large cuttings needed for analysis?);
- bit life;
- performance history;
- previous bit condition.

Figure 1.61. PDC bit.

A typical $12\frac{1}{4}''$ tri-cone bit costs \$10 k, whereas a PDC bit would cost \$50 k and a diamond impregnated bit might cost \$100 k.

Core bits are used for drilling "core" — i.e. a continuous sample of rock. This process is described in the section on Formation Evaluation. The core bit is broadly similar to the diamond bit described above, except that it has a large central hole that "swallows" the core of rock that will be recovered to surface (see Figure 1.131).

Hole openers are a type of bit designed to follow a small hole and open it up to a larger diameter. This is sometimes done because petrophysical logs are of higher quality in the smaller wellbore, whereas a larger wellbore is needed to drill deeper in the formation. A typical hole opener may follow a $12\frac{1}{4}''$ hole and open it up to $26''$ diameter to allow running of $20''$ casing. Figure 1.62 shows a typical hole-opener.

Under-reamers, as shown in Figure 1.63, are similar to hole openers except that the larger size is cut using arms that extend out

Figure 1.62. Hole opener.

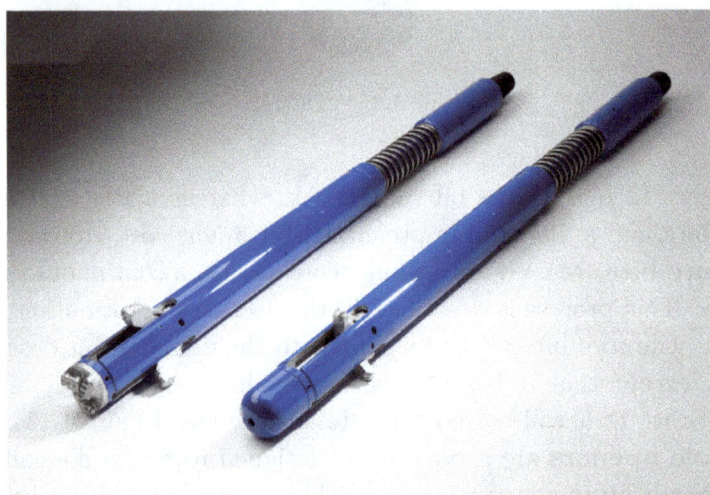

Figure 1.63. Under-reamer.

from the body of the tool. This is done using the hydraulic energy of the drilling fluid being pumped through the tool. The arms are coated with PDC cutters, and nozzles provide fluid to the cutting surface. An advantage of the under-reamer is that it can pass through a smaller casing before the arms are extended, allowing the hole below to be opened to a larger diameter if required. For example, this is a requirement when setting expandable casing. Problems have occurred when the arms do not retract when pumping stops, making recovery of the tool impossible.

1.10. Well Hydrostatics

Understanding hydrostatic pressures in the well is essential for well engineers.

In a static situation, pressure can be expressed in oilfield units thus

$$P = \rho z$$

where:

> P is the pressure downhole (units: psi (Lb/in^2)).
> ρ is the density of the fluid (units: psi/foot).
> z is the vertical depth in the well (units: feet).

Frictional pressure drops in the drill string, across the bit and up the annulus also need to be considered. These are offset by the pressure increase developed at the circulation pump. These effects can be plotted as in Figure 1.64.

In this example, the pump develops 5,000 psi. Assuming a very low flow rate, frictional effects will be almost zero, and with a mud weight of 0.5 psi/ft pressure in the drill string at 10,000 ft will be 10,000 psi. Pressure on the annulus side will be 5,000 psi, giving a 5,000 psi pressure drop across the bit.

Frictional pressure drop always resists flow. If we increase the flow rate, frictional effects become significant and pressure in the drill string at 10,000 ft will **reduce** due to friction in the drill string (in this example by 1,500 psi). On the annulus side, pressure will **increase** due to friction on the annulus side (in this example by 750 psi). The pressure drop across the bit is then reduced (in this example from 5,000 psi to 2,750 psi).

The formation is exposed to the increase in annulus pressure, which can be expressed as **Equivalent Circulating Density (ECD)**. In this example, the ECD is 0.5 psi/ft (mud weight) + 750 psi (friction pressure)/10,000 ft (depth) = 0.575 psi/ft. ECD is

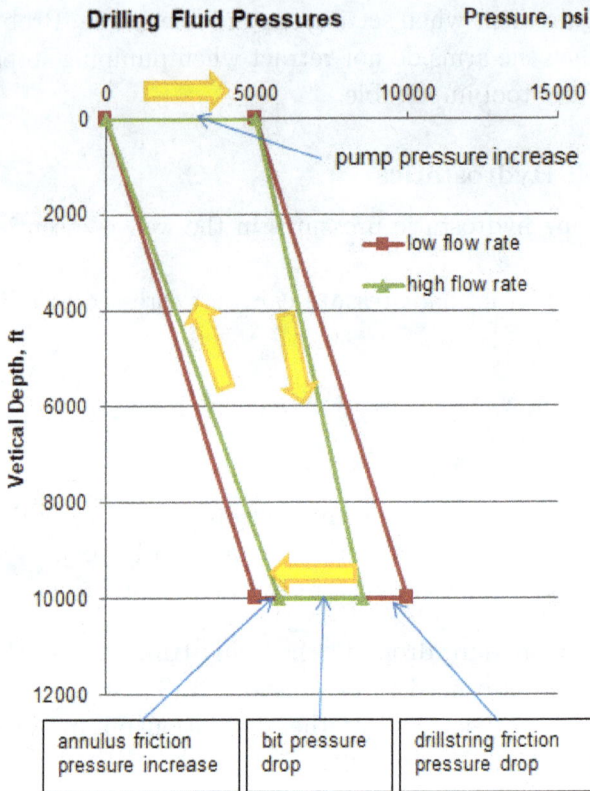

Figure 1.64. Well hydrostatic pressures.

Note: In this example the pump pressure has been kept constant whereas in reality a flow increase would usually result in pump pressure increase. This example really models different bit nozzle sizes.

important because it increases the overbalance against the formation pore pressure, which is often maintained at 200–500 psi. If it becomes excessive, losses into the formation can occur or the formation will fail if the fracture strength is exceeded. These are important considerations especially in HPHT wells where these margins are small. ECD can be minimised by:

- maximising annulus cross-sectional area;
- lower mud viscosity;
- slower pump rates (but keep the hole clean!).

1.11. Drilling of a Basic Well

How is a basic well drilled? At the most fundamental level this is determined by the following criteria:

- depth of the objective;
- diameter of final hole needed;
- formation pore pressure and;
- rock fracture strength.

One might imagine that a well can be drilled to whatever depth is required in just one diameter of borehole. Unless the well is very shallow, this is not possible for the following reasons:

- If gas or other hydrocarbons are encountered, they cannot be contained safely.
- The rocks around the wellbore may swell or collapse, preventing further progress.

On the first point, take a look at Figure 1.65. For a typical oilfield well of depth 11,000 ft this shows how the formation pore pressure and rock fracture strength might increase with depth. It is obvious that if the bottom of the well fills with gas (which has a very low density and hence a near-vertical line on this pressure/depth plot), at some point the formation strength will be exceeded (point A on Figure 1.65). If this condition were allowed to exist, uncontrolled flow could occur from the bottom of the well into the rocks higher up; this would have the potential for contaminating, e.g. shallow aquifer formations and if near the surface, create serious safety issues at the wellsite.

This challenge is addressed by drilling progressively smaller diameter holes as the well is deepened; each hole section is then protected by running casing that is sealed against the rock using cement. This prevents the upper part of the wellbore experiencing pressures that could fracture or cause leakage into the rock. The quality of this bond is tested before drilling ahead the next hole section.

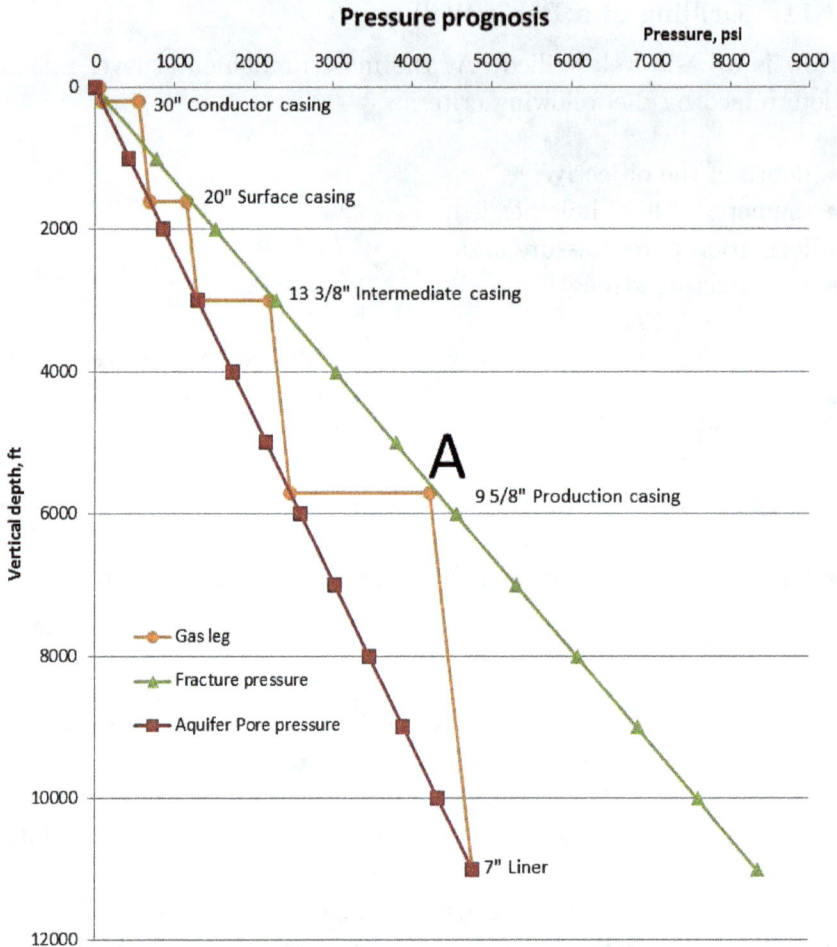

Figure 1.65. Basic setting depths for casing.

Referring to Figure 1.65, again it can be seen that in this (rather simplistic) example, five such hole sections would be needed to reach a total depth at 11,000 ft. At shallow depths, such gas influxes can never be contained. Thankfully, shallow gas is seldom encountered, but if it is, any influx is diverted away from the wellsite using a diverter arrangement (see Section 1.7.4) rather than being closed-in using BOPs.

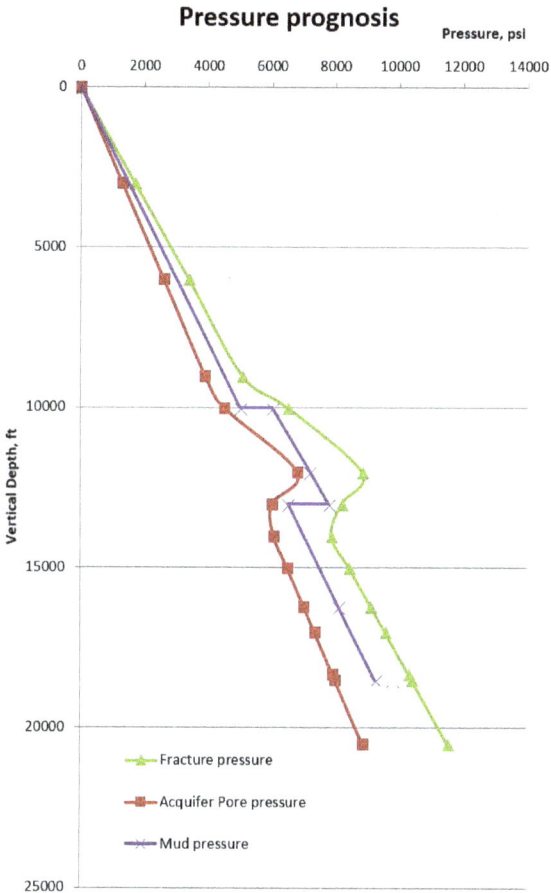

Figure 1.66. Example pore and formation pressures.

A basic requirement for drilling a well is that the hydrostatic pressure exerted by the drilling fluid is greater than the pore pressure in the formation but less than its fracture strength. Figure 1.66 demonstrates a typical example in which the pore pressure gradient and fracture pressure vary with depth. It is clear that step changes in mud density are required to manage this correctly, and this is only possible by setting casing to isolate one section from another.

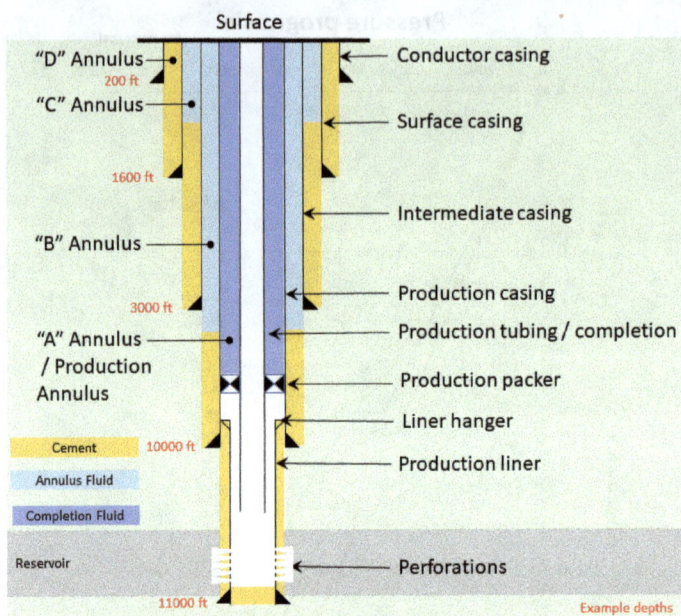

Figure 1.67. Basic casing design and annuli designations.

The "telescopic" nature of the wellbore (see Figure 1.67) — requires that the drill bit used for each section can pass through the casing of the previous section. This creates some rather odd sizes of hole and casing. Given the origin of oilfield drilling in the US, the following imperial sizes are typical.

Name of casing	Outside diameter	Inside diameter	Hole size required
Conductor	30″	28″	36″ (if drilled)
Surface	20″	19.124″	26″
Intermediate	13³/₈″	12.415″	17¹/₂″
Production	9⁵/₈″	8.681″	12¹/₂″
Liner	7″	6.184″	8¹/₂″

The depths at which the various casings might be set are shown in Figure 1.67. This diagram also refers to the naming convention for annuli. Letters are used from the production tubing outwards, so in the example:

Name	Inner string	Outer string
A	Tubing	Production casing/Liner
B	Production casing	Intermediate casing
C	Intermediate casing	Surface casing
D	Surface casing	Conductor

Referring to the basic process steps of planning and constructing a well, the actual drilling activities for the simple well using the casing above would comprise the following:

Section		Depth (ft)	Operation
0	30	200	Drive 30″ conductor
1.1	26″	1,600	Drill 26″ hole
1.2			Run 20″ casing
1.3			Cement 20″ casing
1.4			Nipple up BOPs
1.5			Pressure Test BOPs and 20″ casing
2.1	17½″	3,000	Drill out 20″ casing shoe
2.2			Formation Integrity Test
2.3			Drill 17½″ hole
2.4			Run 13⅜″ casing
2.5			Cement 13⅜″ casing
2.6			Pressure Test BOPs and 13⅜″ casing
3.1	12¼″	10,000	Drill out 13⅜″ casing shoe
3.2			Formation Integrity Test
3.3			Drill 12¼″ hole

(Continued)

(Continued)

Section	Depth (ft)	Operation
3.4		Run 9⅝″ casing
3.5		Cement 9⅝″ casing
3.6		Pressure Test BOPs and 9⅝″ casing
4.1	8½″ 11,000	Drill out 9⅝″ casing shoe
4.2		Formation Integrity Test
4.3		Drill 8½″ hole
4.4		Run 7″ liner
4.5		Cement 7″ liner
4.6		Pressure Test BOPs and 7″ liner
5.1	Completion	Run completion

1.12. Casing and Casing Design

Casing purpose. The purpose of casing (together with cement) can be summarised as follows:

- **Support the wellhead** — to take the loads on the wellhead during drilling and when the well is operating. For example, on an offshore platform the conductor is designed to withstand environmental side-loads (e.g. wind, waves and tides);
- **Provide wellbore stability** — by physically supporting the formation and preventing intrusions (such as salt squeezing) into the wellbore;
- **Isolate different formations** — required to prevent contamination between formations such as drinking water aquifers;
- **Contain well pressures** — act as a pressure-containing vessel (with BOP) during drilling, production and intervention;
- **Isolate loss zones** — to avoid drilling problems.

Casing design is a complex subject, but some basic principles are covered here. Casing can be considered as a very long and thin pressure vessel run into a hole in the ground and cemented in place. As such, the pressure vessel has to meet several design criteria.

Burst pressure. During several phases of the well life, pressure inside the casing will exceed that on the outside. This will occur, for example, in drilling if the casing fills with gas from a lower part of the well and in the production phase if a leak occurs in the production tubing. A typical criteria is that casing needs to be able to withstand a gas gradient from the bottom of the next hole section or from the reservoir in the case of production casing. If a well is to be used for gas- or water-injection, surface injection pressures will be even greater than reservoir pressure, and so the casing scheme will need to be designed accordingly.

Collapse pressure. In other phases of well construction and operation, pressure on the outside of the casing will be greater than that on the inside. This will occur, for example, during the drilling phase if the well is not kept full of liquid due to drilling fluid losses in a lower hole section, and when cementing the casing in place.

Tension/compression/bending. When hanging vertically, the casing will be in tension due to its own weight in air. As it is run into the hole, there will be a buoyancy effect due to displaced drilling fluid. If the hole is not vertical and curved, side-loads and bending moments will be applied. Shock loads as the casing is landed off on the rig floor need to be factored in. Friction in the well will reduce tension when the casing is being run, but increase it if the casing has to be retrieved. An important criterion is that the rating of the rig being used must be sufficient to support the casing when it is being run, and pull it back should that be needed.

During the design phase, it is normal to model a series of load cases for a specific well using software. This typically includes tri-axial analysis that combines all of the loads mentioned above. Usually this modelling takes place on a per-casing basis, but more sophisticated tools are available for analysis of multiple casing strings.

Once the load cases are complete, they can be compared with the strength capability of various casing designs. Typically the overall size of the casing will be determined by the hole diameter, previous casing internal diameter and the various radial clearances needed for circulation of drilling fluids and cement. The wall thickness of the casing (often expressed as the linear unit weight in lb/ft) and the material specification (in terms of steel yield strength) can be chosen from a list of industry standard products to meet the load cases. Take for example the following specifications for various types of $9^5/_8''$ casing:

Weight (Lb/ft)	Wall thickness (in)	Grade	Burst (psi)	Collapse (psi)	Tension Lb × 1,000
47	0.472	L-80	6,870	7,100	1,086
47	0.472	C-95	8,150	5,090	1,289
47	0.472	P-110	9,440	5,300	1,493
53.5	0.545	L-80	7,930	6,620	1,244
53.5	0.545	C-95	9,410	7,340	1,458
53.5	0.545	P-110	10,900	7,950	1,710

Burst, collapse and tension ratings all generally increase with greater wall thickness (linear weight) and higher material yield strength as would be expected. There are other criteria to be considered however. These include the strength of the casing threaded connections, which usually but not always, exceed the casing body strength. Some casing grades — typically those of higher yield strength — cannot tolerate H_2S gas and consequently cannot be used in wells where sour gas may be expected.

Essentially, casing design is also determined by economics — stronger casings are more expensive, and hence a casing string that *just exceeds* requirements is usually chosen. It is not practical to store and transport every single option from a casing supplier's catalogue — hence the choice is usually made from a reduced list.

Sometimes, a combination of casing weights and grades is used in one string, as the load criteria varies with depth.

Connection design is also very important, and over the history of the oilfield there have been significant developments. Casing connections include a collar of diameter greater than the casing, into which the casing pin threads are screwed — (see Figure 1.68). The most basic designs do not seal against gas flow, may not be as strong as the pipe body but are inexpensive to machine on the end of the pipe. A more sophisticated connection exceeds pipe body strength, is gas-tight and may allow casing rotation, e.g. for drilling with casing applications. However, this type of connection will be proprietary to a certain casing manufacturer or supplier company and more expensive. Casing pipe grease is used to lubricate the connection, but the actual seal is created by the steel geometry of the connection. When first developed, a great deal of testing (called qualification) of casing threads takes place to prove their capability and reliability to meet the specifications at various downhole temperatures. The make-up torque of the connection is also tightly specified and controlled during running operations with tongs instrumented to measure, check and record rotation and torque applied. Casing threads are also inspected on the wellsite prior to running.

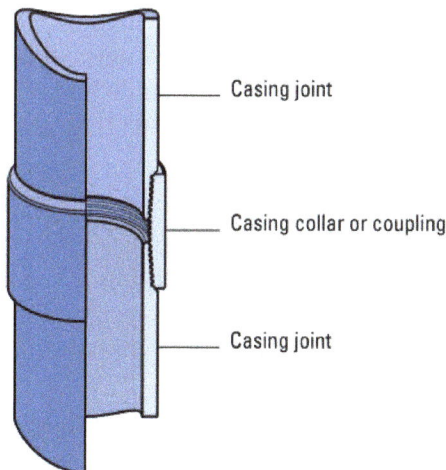

Figure 1.68. Casing connection.

When drilling occurs through a previous casing, wear will take place. Although this is minimised, it will result in local reduction in wall thickness. There may also need to be an allowance for some corrosion of the casing over its lifetime. These and other uncertainties in possible conditions downhole are compensated for by using a Safety Factor in some casing loads — typically 1.1 for burst and 1.3 for tension ratings.

At the top of the casing is a casing hanger which screws on to the last joint of casing and provides a shoulder to support the weight of the casing in the wellhead.

A liner is very similar to a casing except that it is hung off inside a previous casing rather than from a wellhead, using a liner hanger. The design considerations are also very similar to those for casing. The liner hanger usually allows the liner to be **tied-back** to surface — i.e. an additional casing run from surface after the liner has been cemented that seals in the top of the liner hanger. The tie-back string is usually the same diameter as the liner but is not cemented in place. The tie-back packer has a Polished–Bore Receptacle (PBR) that allows axial movement of the tie-back string while retaining pressure integrity of the full string. Some operators believe that this arrangement results in a better liner cementation because less backpressure is applied during the operation, minimising losses.

Expandable tubulars are a recent development. The expandable casing is run into a well and expanded *in situ* up against the borehole wall (see Figure 1.69) using a cone that is pulled through the casing string and anchored inside the string further up. The outside diameter can be increased by up to 25%, and because it presses up tightly against the borehole wall it does not need to be cemented. The technique is useful if a **contingency** casing is required due to wellbore stability issues or weak formation strength. Because the casing can be run through an existing string of similar diameter (but subsequently expanded), in theory a monobore well could be drilled from surface to Total Depth (TD) using a series of expandable casings.

Components in the casing and liner that are used for cementing are covered in the cementing section (Section 1.16).

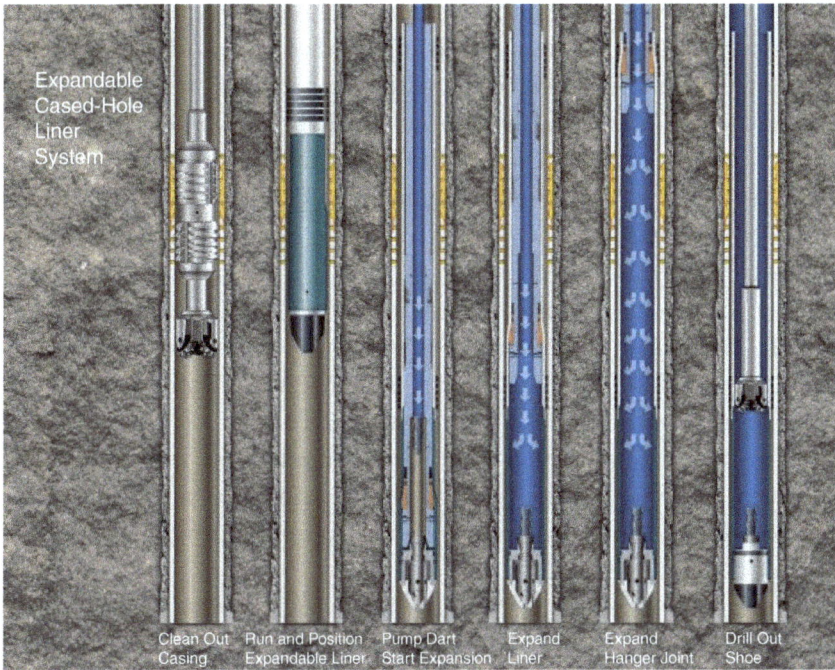

Figure 1.69. Expandable casing.

1.13. Pressure Testing

Pressure testing is an essential part of safe well construction, to confirm that parts of the well are capable of retaining pressure. The following are the main examples of pressure tests:

- BOPs and associated equipment;
- casing after cementation;
- formation strength prior to drilling a new hole section;
- completion equipment;
- inflow tests.

It is a mandatory requirement in most jurisdictions and certainly part of a typical operator's standards that the BOP be pressure tested on a regular basis, often every 14 or 21 days. This includes choke and kill lines, manifolds, IBOPs and other relevant equipment. It is normal to perform tests at low pressures and at the maximum working pressure of the equipment, for 15 min per item. All remote equipment is also function-tested — i.e. actuated to ensure that it is working correctly. If an item fails the test, it must be repaired or replaced before commencing operations. Records must be kept and made available for inspection if requested. Testing BOPs is time-consuming (especially for subsea BOPs) and requires pulling the drill string. Special tools are used to seal the BOP off from the rest of the well during the test.

The casing is normally pressure tested immediately after displacing the cement. This is so that any microannulus created by the casing radially expanding during the test will be sealed by the cement as it sets. Typically, casing is tested to the highest expected pressure it will experience, not necessarily its full rated pressure. Pressure tests are generally carried out using the cementing unit, and volumes pumped for the test are carefully monitored and checked against theoretical values.

Formation strength tests (also called leakoff tests or Formation Integrity Tests (FITs)) are usually carried out before drilling ahead in a new hole section. The purpose is to confirm the formation strength which is an important variable for well control planning. The test is carried out after drilling a few feet of new hole and circulating the well to assure a uniform and known mud gradient. Pressure is applied gradually to the wellbore by using the cement unit and closing the BOPs. The pressure is plotted against volume until the Formation Fracture Pressure (FPP) is attained (see Figure 1.70). The formation strength gradient can then be calculated as the mud gradient + (FPP divided by the true vertical depth of open hole).

Completion equipment is tested in a very similar way to casing. Often, wireline plugs are set and used to test against. Sequential tests are used so that if remedial action is required this is identified as soon as possible to mitigate wasted activities. For example, a liner might be pressure tested, the production packer, the tubing, the subsurface safety valve (and control lines), the Christmas tree etc.

Inflow tests (sometimes called negative pressure tests) are required to assure that flow (usually from the reservoir) cannot enter the wellbore. In this case, the pressure in the well is reduced, usually

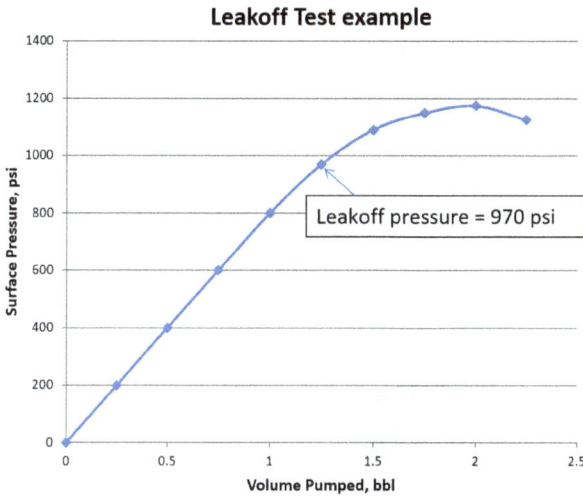

Figure 1.70. Typical leakoff test.

by displacing to a lighter drilling fluid or water and reducing surface pressure in stages. The well is then monitored for flow over a period of 30 or 60 min, and if none then pressure integrity is confirmed.

1.14. Wellheads and Christmas Trees

1.14.1. *Wellheads*

The purpose of the wellhead is as follows:

- provides a pressure-containment interface with BOPs during the drilling phase;
- provides casing hang-off profiles during the well construction phase;
- provides a hang-off profile for production tubing;
- allows access to the casing annuli;
- supports the Christmas tree;
- in combination with Christmas tree, provides surface flow-control during the production phase.

Each casing string is landed off in the wellhead, which supports the vertical load of the casing and forms a seal around the casing hanger. Sometimes, **tie-down** bolts are used to lock down the hanger; alternatively, a separate **seal assembly** is used to lock down the hanger and seal against the wellhead.

In the simplest of wellheads, a spool is added after cementing each casing string that seals against the previous one and into which the next casing is landed (see Figure 1.71). Access to each annulus

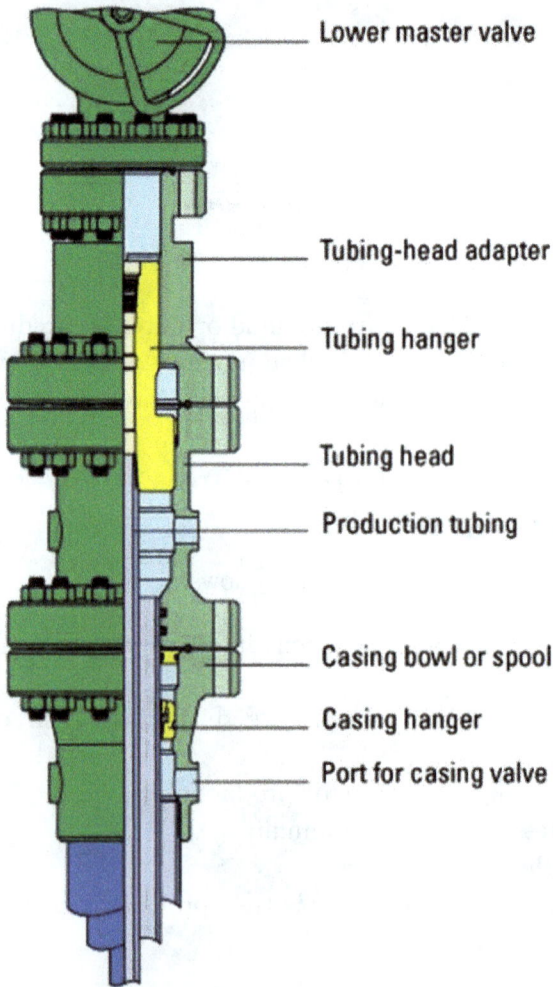

Figure 1.71. Traditional wellhead.

between casings is achieved with a side-outlet port on to which a valve is normally fitted. The production tubing is also usually suspended in the wellhead in a similar manner to casing strings. In some cases, the hanger is locked-down in the wellhead so that pressure from below cannot push the hanger upwards in the wellhead.

A downside of the basic spool-type wellhead is that the BOP must be temporarily removed after cementing the casing to add another spool, during which time there is no primary well control barrier. This can be avoided using a compact wellhead in which casings and tubing can be hung off in a single unit (see Figure 1.72). This also has benefits in terms of saving time and space required (important in an offshore environment).

Subsea wellheads are a variation on the compact wellhead but are designed to work without manual intervention as this is not possible in deepwater (see Figure 1.73). Up to five or six casings can be hung

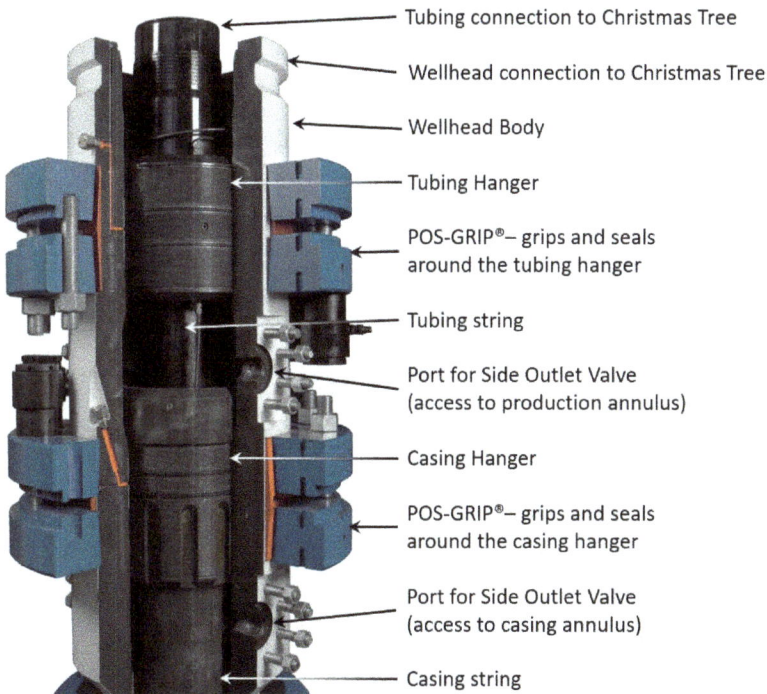

Figure 1.72. Mono-block wellhead. Copyright Plexus Holdings plc.

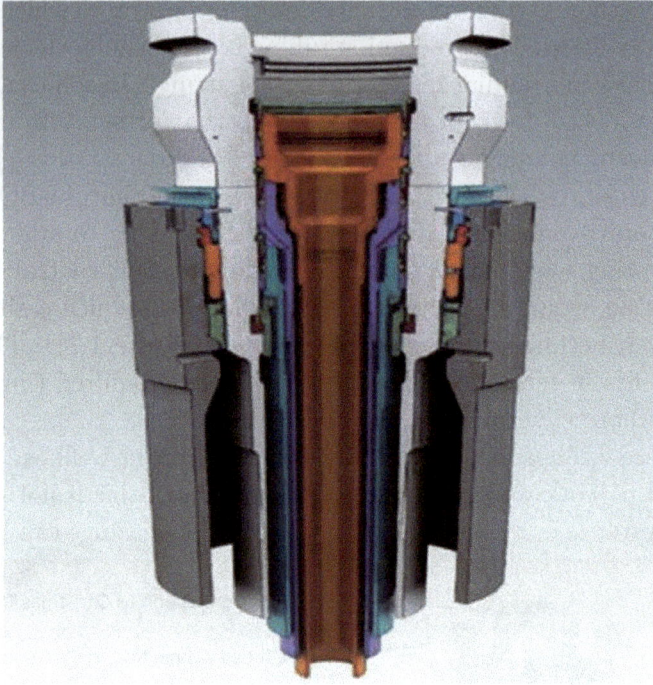

Figure 1.73. Subsea wellhead.

off in a typical subsea wellhead. No subsea wellhead systems allow access to the inter-casing annuli (i.e. with the exception of the A- or production annulus). This lack of access has implications for casing design and how the well is drilled and operated during the production phase.

1.14.2. *Christmas Tree*

The Christmas tree:

- Is installed on top of the wellhead to control the flow of well fluids during production.
- Provides primary and back-up control facilities for production.
- Enables wellbore shut-in.
- Incorporates facilities to enable safe access for well intervention operations, e.g. slick-line, electric wireline and coiled tubing.

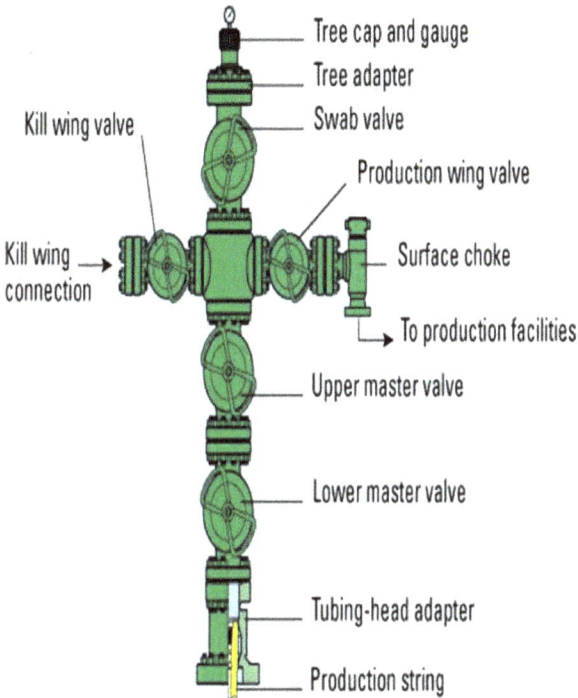

Figure 1.74. Conventional Christmas tree.

Historically the Christmas tree has comprised a series of gate valves arranged in a cross (see Figure 1.74). The lowermost component is an adapter spool that seals against the wellhead and the upper part of the production tubing hanger. The two Master valves (called **Upper Master Valve (UMV)** and **Lower Master Valve (LMV)**) are the main ways of closing in the well. Typically, the LMV remains open as a backup and normal activities are managed with the UMV — sometimes this is activated remotely. Above the UMV is a flow cross to which three other valves are connected. Under normal circumstances, production from the well flows through the **Production Wing Valve (PWV)** and a surface choke to the production facilities. All the valves described thus far are designed to be either open or closed. The choke is used at any intermediate setting during the production phase to control the **Tubing Head**

Pressure (THP) and the flow-rate from the well. The **Swab Valve (SV)** is used to control vertical access to the well should this be needed — e.g. if wireline logging tools need to be run in the well. The **Kill Wing Valve (KWV)** is the final connection and is used in case high-density (kill) fluid needs to be pumped into the well — if necessary at short notice that may preclude disconnecting the main flow line.

On modern land and offshore platform installations, a unitary Christmas tree arrangement is used in which all valves sit in a single steel block. There are also typically several pressure and temperature sensors on the Christmas tree, plus the option to inject chemicals either into the Christmas Tree itself or downhole. Pressure-tight connections are also provided for hydraulic control lines, power and instruments for downhole completion components. A typical setup on land is shown in Figure 1.75.

Subsea Christmas trees are designed to be operated without use of divers. On subsea trees, all the main valves are hydraulically

Figure 1.75. Unitary wellhead and Christmas tree hooked-up to flowline.

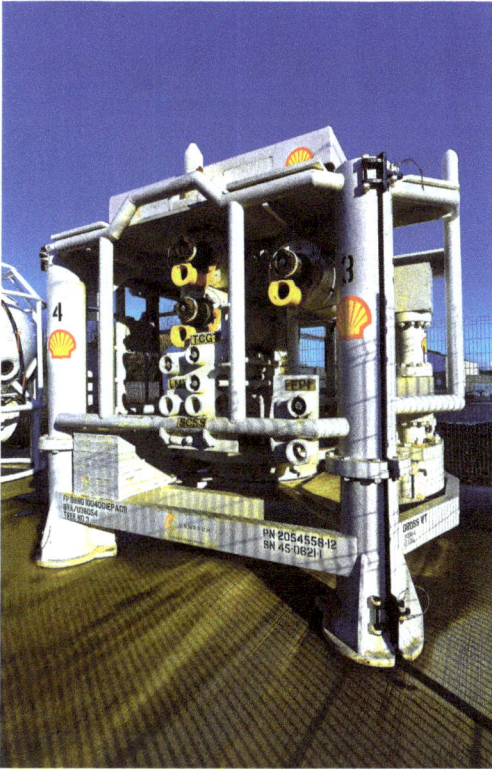

Figure 1.76. Subsea Christmas tree.

operated, frequently with ability to operate in an emergency using an ROV (see Figure 1.76).

A typical "work-class" ROV is shown in Figure 1.77. These devices are "unmanned submarines" that are launched as required from the drilling rig or platform, in a cage that is lowered over the side or through the moon-pool of a rig. The ROV then "swims" from the cage connected to a tether that provides power, control and data communications. The simplest ROVs are used just for observation using a TV camera. The more sophisticated versions (which are the size of a large car) are equipped with several manipulator arms controlled from the surface to carry out activities such as operating valves, collecting samples, picking-up small objects, etc.

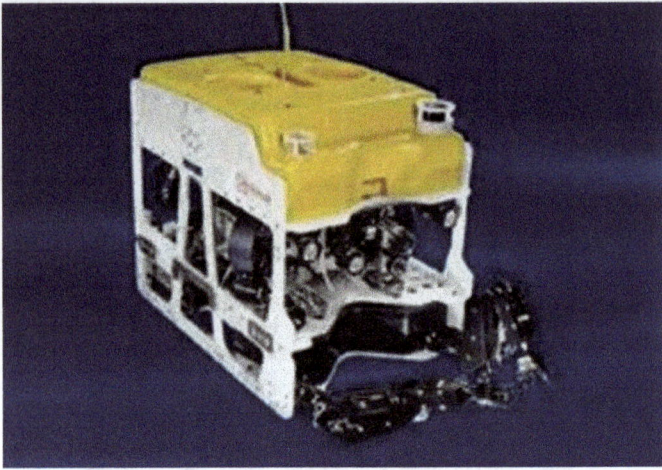

Figure 1.77. Seaeye Cougar 3000m ROV (Remote Operating Vehicle).

Like casing, wellheads and Christmas trees are designed for certain pressures (typically 3,000, 5,000, 10,000 or 15,000 psi), and for normal or sour service. The top flange of the wellhead connects with the BOP during drilling operations and with the Christmas tree during the production phase.

1.15. Completion Equipment

Once the well has been drilled and the casings run and cemented, a final set of tubulars are run into the well to provide a conduit for oil or gas to flow to surface (the situation is similar in the case of a water or gas injection well). This innermost tubular string is called the "**completion**" and comprises tubing and several other components that are described in this section. There are many different types of completion, so this section only provides a general summary of typical arrangements.

The completion enables safe and efficient control of fluid production from the reservoir to surface.

The completion design is closely allied with the type of lift used to help the well flow. Naturally flowing wells are relatively simple; wells requiring pumping or gas-lift require further equipment.

There are other ways of categorising completions — such as the following:

- Cased-hole: The hydrocarbons flow through perforations (holes) in the casing. This has the advantage that unwanted flow (for example water in an oil well) can be controlled.
- Open-hole: No casing is run across the reservoir, and the open formation is allowed to produce freely.
- Gravel-packed: Gravel is required in the lower completion to prevent sand production.
- Multiple zones/strings: Production may be required from multiple zones, sometimes up different tubing strings (i.e. single, dual or triple completions).
- Smart: Downhole control of flow from various zones using Inflow Control Valves (ICVs).
- Pressure/Temperature/Flow data: can be provided from single or multiple sensors in the well.
- Requirement to inject chemicals into the produced fluid to prevent corrosion, reduce wax formation or oil viscosity.
- Need for future treatment such as fraccing, stimulation, reperforation and future logging or other data acquisition.

Unlike casing which is run for the lifetime of the well, completions may be pulled and rerun, though this is avoided if possible.

Figure 1.78 illustrates a typical completion arrangement that will be used to explain the various components.

1.15.1. *Tubing*

The tubing is hung off the wellhead as already explained above. The tubing itself is very similar to small diameter casing and is designed to optimally lift the hydrocarbons out of the well (this is covered in the Production Technology section of this Handbook). Typical tubing diameters are $3^1/_2''$, $4''$ and $5''$. Tubing also has to withstand burst, collapse, tension and sometimes compression loads, for which detailed design is required along similar lines to that of casing.

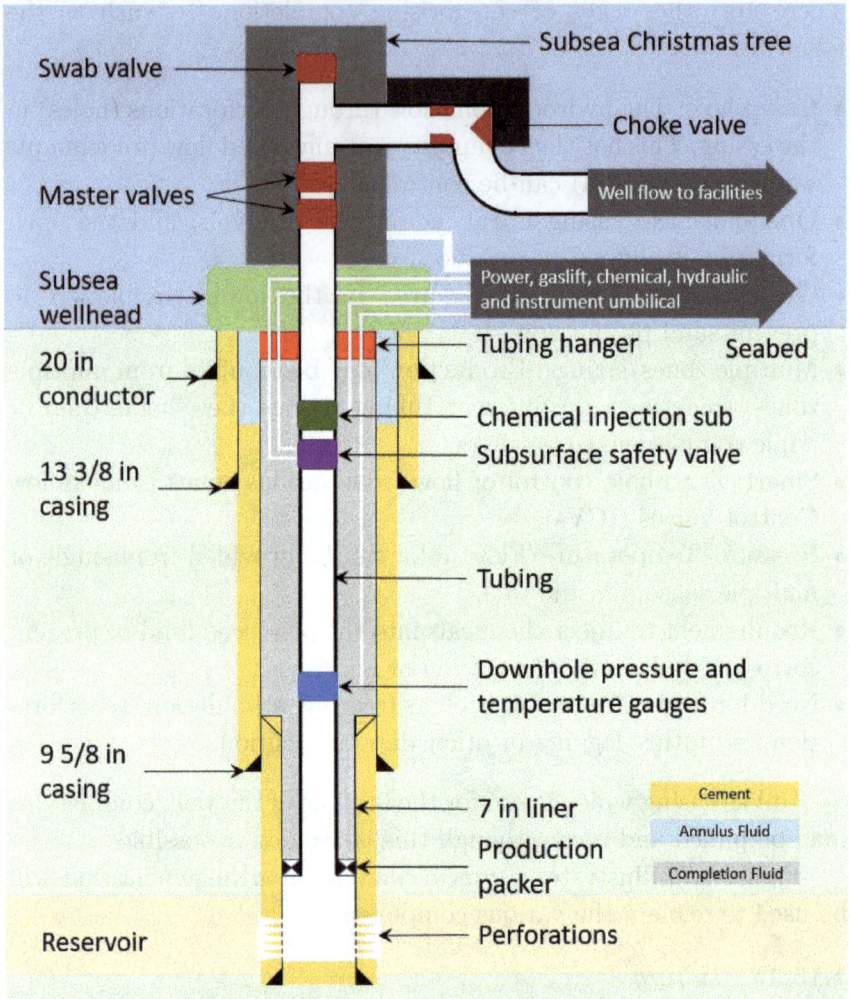

Figure 1.78. Typical subsea completion arrangement.

1.15.2. *Tubing Hanger*

The tubing hanger is similar to the casing hanger except that it seals also into the base of the Christmas tree and often includes one or more ports through which hydraulic and/or electrical control systems

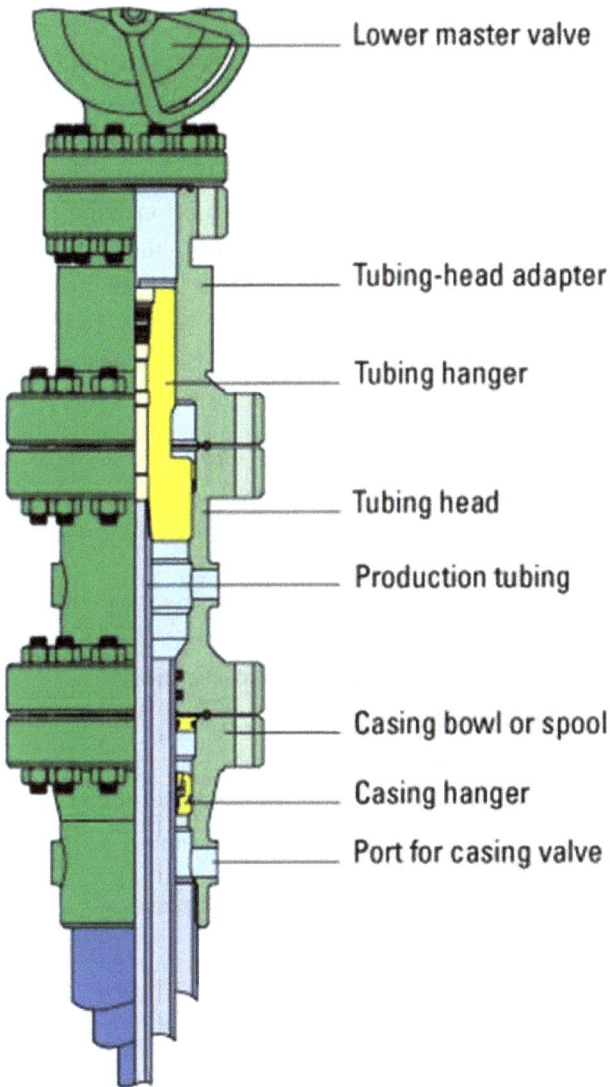

Figure 1.79. Tubing hanger in wellhead/Christmas tree assembly.

pass and are sealed. The tubing hanger also contains a profile that allows a wireline plug to be set to seal off the well if required (see Figure 1.79).

1.15.3. *Packer*

At the lower end of the tubing, a packer is often used to seal between the tubing and the casing. The seal created by the packer creates the possibility of circulating fluid into the well to over-balance reservoir pressure and kill the well, should this be required. Packers are also used to isolate one production zone from another. There are many different types of packers, but typically all have elastomer sealing elements and one of more sets of slip elements that physically anchor the packer and tubing to the inner wall of the production casing (see Figures 1.80 and 1.81). Packers are typically set in place using pressure in the tubing, or using a setting tool run on wireline which is operated with an explosive charge. Some can

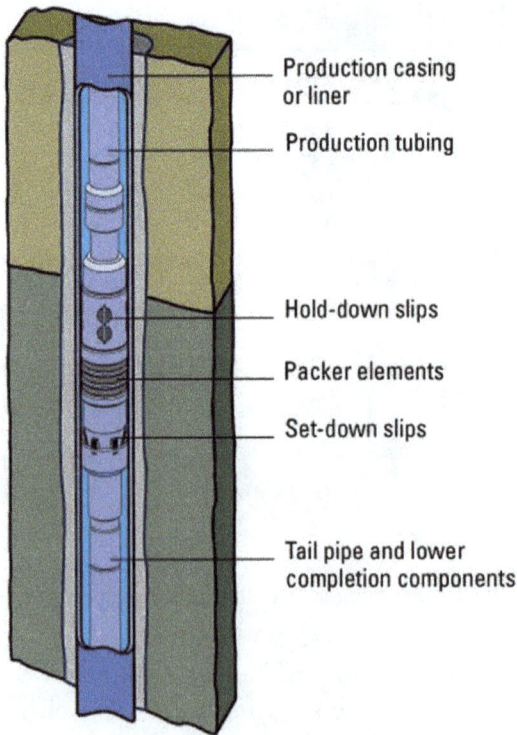

Figure 1.80. Completion packer.

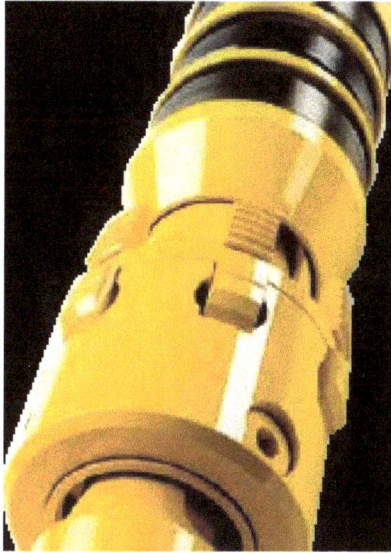

Figure 1.81. Packer.

be released using tubing pull or by explosive cutting of an inner mandrel. They may include ports for power, signal or hydraulic connections.

1.15.4. *SubSurface Safety Valve*

In the event of a catastrophic failure of the wellhead and/or Christmas tree, many wells would flow hydrocarbons to the surface, resulting in environmental damage and possible injury to those at the wellsite. The SubSurface Safety Valve (SSSV) is designed to close off this flow in such an event. It is usually placed in the completion string about 100–500 m below the wellhead, and when the well is operating normally it is open. It is controlled from the surface using hydraulic pressure in a control line which runs alongside the tubing and passes through the tubing hanger and surface equipment to a control system. If the pressure is released — either intentionally or in the event of wellhead failure — the valve (either a ball or a flapper) closes under the action of a spring (see Figure 1.82). Many SSSVs

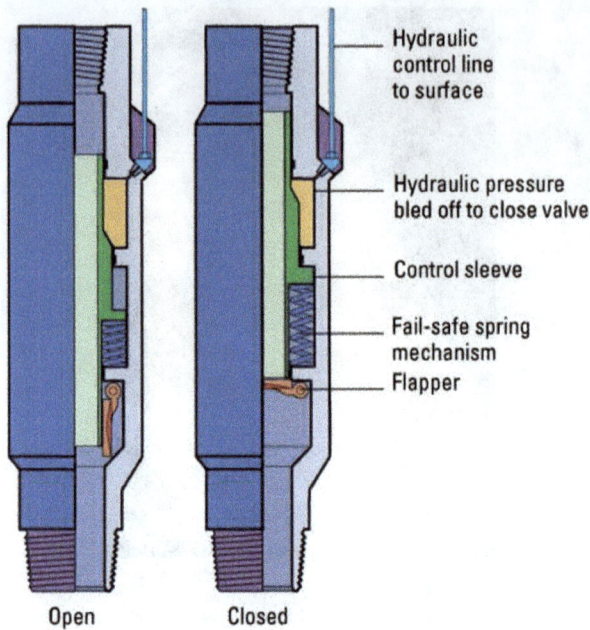

Figure 1.82. SubSurface Safety Valve (SSSV).

also allow an "insert SSSV" to be installed on wireline above the permanent one in the event of failure of the main flapper valve. SSSVs are not used in all wells — e.g. on remote land wells where the probability and/or consequences of well failure are lower and the economic costs more difficult to justify.

1.15.5.   *Sliding Side Door*

Under normal operations, the tubing is isolated from the production annulus (between tubing and production casing). The Sliding Side Door (SSD) is a component that allows communication between tubing and annulus when required. It comprises a sliding sleeve that uncovers holes in the tool body to allow flow from tubing to annulus (see Figure 1.83). The sliding sleeve is operated by a wireline tool (wireline operations are covered in more detail later in the chapter), or in more sophisticated versions by remote hydraulic or electrical

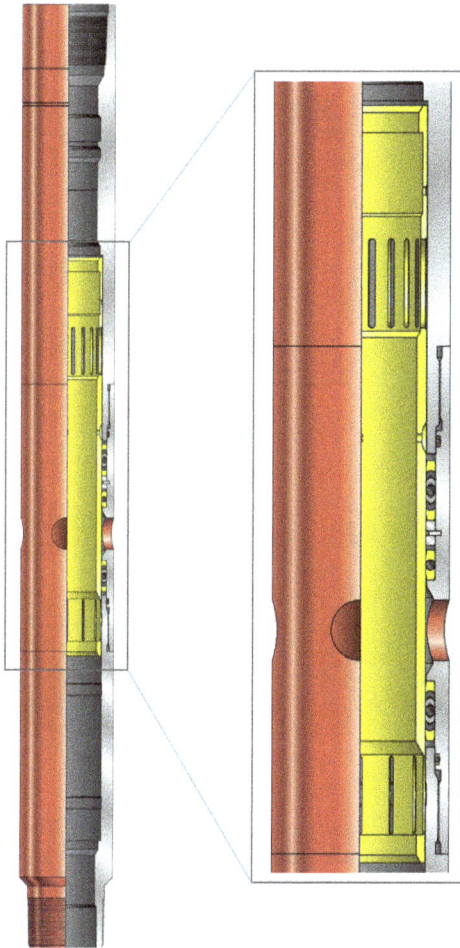

Figure 1.83. Sliding Side Door (SSD).

actuation. SSDs can also be installed opposite different completion zones to allow selective production from each.

1.15.6. *Plug Nipple*

This is a simple component — with no moving parts — run in most completions that allows the setting of a plug in the tubing to isolate part of the completion and prevent flow. The plug is latched into an internal profile that physically prevents the plug moving up or down

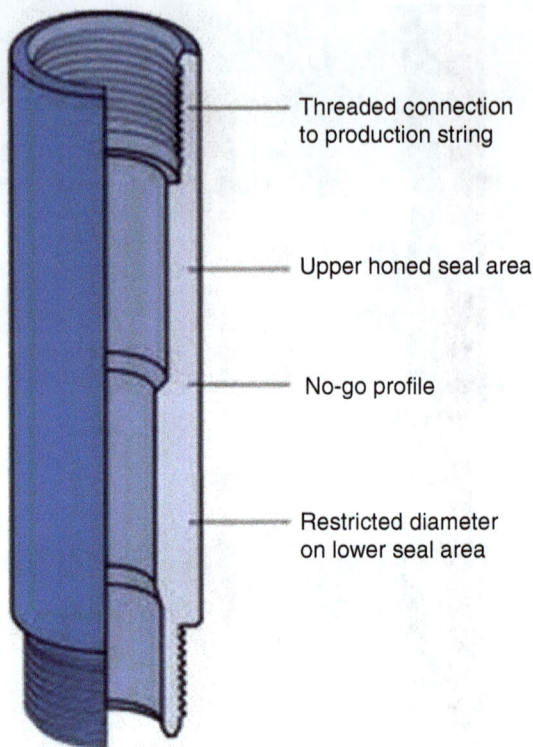

Figure 1.84. Plug nipple.

the tubing due to pressure, and a smooth bore area against which elastomeric elements on the plug can seal. The latching profile in the nipple allows selective setting of plugs — i.e. so that a plug can only be set in the correct nipple rather than one above or below (see Figures 1.84, 1.85 and 1.86).

1.15.7. *Telejoints*

Most completion strings are in tension when landed and set. During production operations, the tubing heats up and tension in the completion string is reduced. If this were to result in compression in the tubing string, there is a risk of buckling and tubing damage.

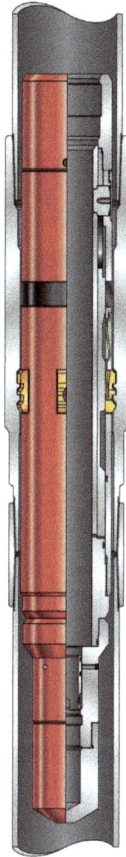

Figure 1.85. Plug set in nipple.

This can be prevented by running a telescopic unit — the Telejoint — to accommodate such forces.

1.15.8. *Gas Lift/Side Pocket Mandrel*

Where a well is not able to flow naturally (e.g. reservoir pressure is low), the density of produced fluid in the well can be reduced by injecting gas into the tubing. This is done by filling the annulus with gas and using a Side Pocket Mandrel (SPM) in the tubing string. The

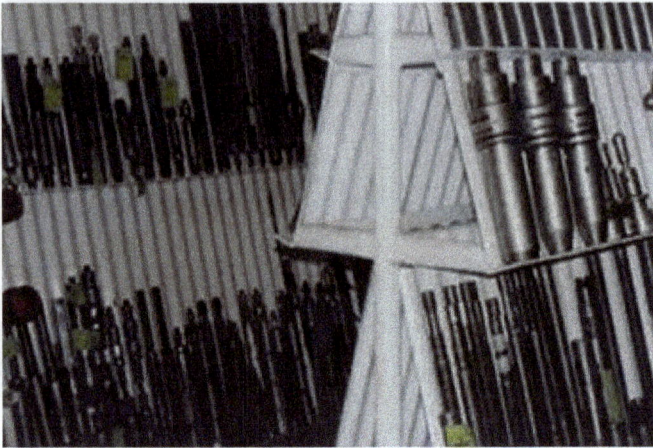

Figure 1.86. Typical store of slick-line plugs.

gas enters the tubing via a pressure-operated valve set into the SPM by a wireline activated tool called a kick-over tool. In most designs, several SPMs are installed at various depths in the well between tubing hanger and packer. The gas lift valves in these SPMs open sequentially when the well is first kicked-off, so that in the operated mode only the deepest is open. If required, SPMs are also used for the injection of chemicals. Physically, the SPM has a bulbous section (the "side pocket"), which has ports to the annulus and a sealing area (rather like the SSD above). In the main bore above the side pocket there is an orientation arrangement so that the gas lift valve running tool can locate and "kick-over" into the pocket (see Figures 1.87 and 1.88).

1.15.9. *Sensors*

Obtaining data on downhole conditions strongly supports better reservoir understanding and hence management and control, as well as well integrity. Zones can be reconfigured either remotely or by intervention and recompletion. If more detailed information is required, this can be obtained using wireline logging techniques, but in some (e.g. subsea) wells the cost of accessing the well can be prohibitive.

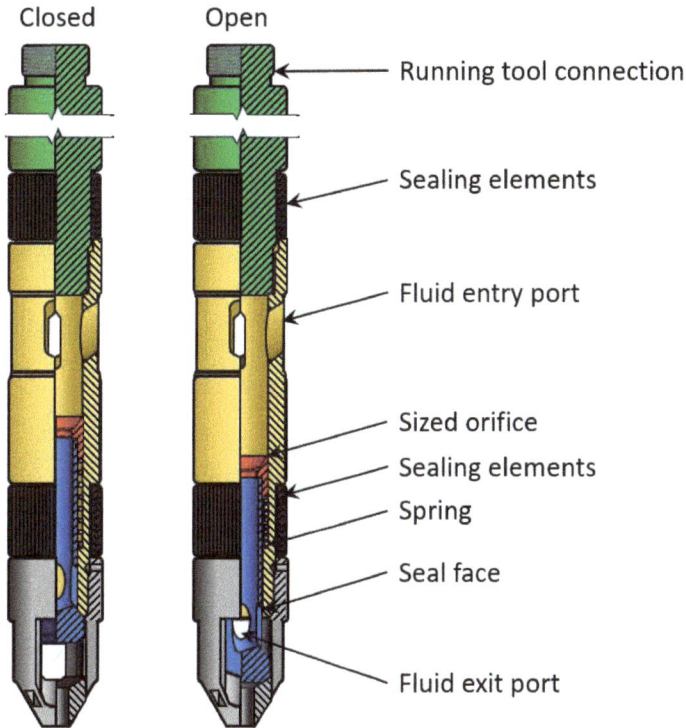

Figure 1.87. Gaslift valve.

Modern sensors measure pressure, temperature, flowrate, vibration, fluid type and other parameters. High accuracy and reliability are essential in these applications, and over recent years the reliability and scope of these devices has increased substantially. Data transmission takes place via electrical cable, fibre-optic cable or by electro-magnetic transmission up the pipe. Some devices include backup memory in case of transmission failure. Downhole power can be provided by long-life batteries or by power generation devices set in the fluid flow.

1.15.10. *Signal and Control Lines*

Several examples of signal and control cables have been described above. There may also be electrical power lines run to operate

Figure 1.88. Side pocket mandrel in completion.

downhole artificial lift devices such as Electrical Submersible Pumps (ESPs, see Section 1.31 on Artificial Lift). These cables are run alongside and strapped to the tubing and other components in the well, and through seals in packers and tubing hanger. Their reliability

Figure 1.89. Wire-wrapped screen.

is paramount, and great care is taken when running the completion to ensure that these cables are not damaged.

1.15.11. *Gravel Packs*

Some formations produce sand along with oil or gas — which can start or increase as the reservoir pressure is reduced or water is drawn into the well during its life. Sand can do a great deal of damage to surface facilities such as abrasively wearing away pipelines and blocking facilities. An effective way of managing sand is to block it off at the reservoir using a gravel pack. The basic technique is to run a "sand screen" across the production interval comprising solid pipe with wire wrapped around it (see Figure 1.89). Then, between the formation and the screen, gravel is placed. This gravel is uniform and sized to prevent formation sand being produced but allowing fluids to flow through. Alternatives include external gravel packs — where the gravel is placed directly against the formation — and internal gravel packs where the gravel and screen

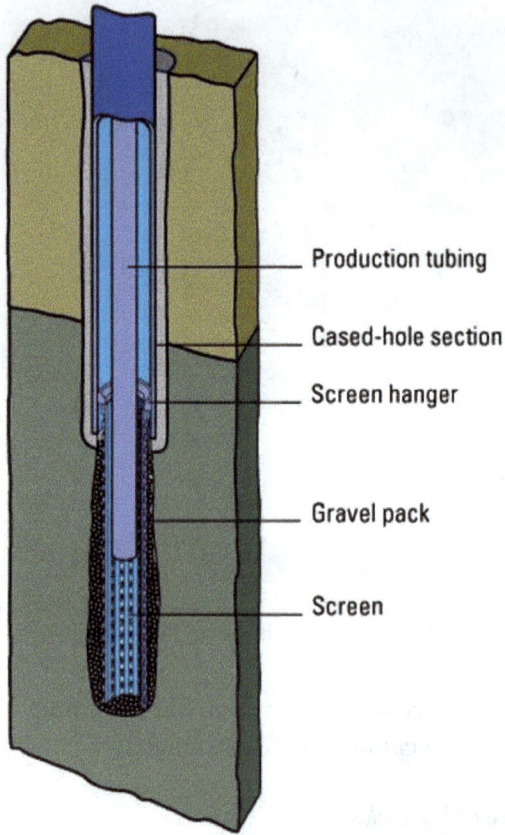

Figure 1.90. External gravel pack.

are inside a perforated liner or casing. The gravel is "placed" in the well using specialist techniques — essentially similar to a cementing operation but using sand suspended in a gelled liquid (see Figure 1.90).

1.15.12. *SMART Wells*

Wells that include downhole sensors and control systems are sometimes called smart wells. Several trends are driving this. To optimise ultimate recovery from the reservoir, there is a need to collect data and to use this to manage production from each zone more proactively and deliberately than before. The required levels of measurement, processing and physical control are increasingly possible with new technology that is more capable, more reliable and of lower cost. Alternatives such as workover and re-entry have become increasingly expensive and difficult to schedule without interfering with other activities including production.

1.15.13. *Swellable Elastomers*

Swellable elastomers are recent inventions with the potential to replace passive sealing elements in certain components. These elastomers can be formulated to react either to water or to oil and swell to many times their original volume. This can be useful for shutting off water, for example, in wells where water might be expected later in field life.

1.15.14. *Complexity, Reliability and Maintainability*

In all the above completion equipment, a natural trade-off exists between completion reliability and complexity, which needs to be resolved during the design phase of the well. Due consideration must also be given to how the well will be maintained over its lifetime which may span 25–50 years.

1.16. Cementing

Cementing of the casing is the final crucial operation in securing a well section. Cement is a barrier to flow in the well and the following must be achieved for it to be effective:

- high-quality uncontaminated cement that bonds to casing and formation;
- high compressive strength;
- well control throughout the cementing operation;
- reliable float equipment (to prevent flow inside casing);

- no channels in the cement (to prevent flow outside casing);
- isolation of hydrocarbon zones;
- isolation of water zones;
- protection of the casing from corrosive fluids.

A good cementation is achieved through:

- pre-cementation circulation and conditioning of mud;
- proper mixing and blending of mixwater and cement, e.g. to minimise free water and to achieve the correct density;
- high displacement rates;
- density difference between spacer, lead, main & tail slurry;
- effective mud removal;
- casing movement (reciprocation and/or rotation);
- casing centralisation;
- managing risks such as fluid losses and not getting the casing to the bottom of the well.

1.16.1.   *Oilfield Cement*

Oilfield cement is specifically manufactured for use downhole, as specified in API standard 10A. It is available in variety of "classes" appropriate to requirements — A, B, C, D, E, F, G and H. Cement is modified with chemical "additives" to deliver required properties, for example:

- set quicker (accelerators) or slower (retarders);
- increase or reduce slurry density;
- reduce setting-time sensitivity to temperature;
- increase compressive strength or elasticity;
- possess suitable flow properties — PV/ YP and gels.

1.16.2.   *Basic Casing Cementing Process*

The following are the basic steps of casing cementation:

	Operation	Comments
1	Plan operation, check people, equipment and chemicals Take and test samples	Proper planning essential
2	Drill hole	
3	Circulate well clean of cuttings	To ensure good cementation
4	Pull out drill string and BHA	Check that drilled hole is ready for casing
5	Run casing, while filling with drilling fluid	
6	Rig-up systems to cement casing	
7	Circulate drilling fluid down casing up annulus	
8	Pump spacer ahead	To separate cement from drilling fluid
9	Drop bottom cement plug	
10	Mix and pump cement	Check slurry density, pump rate, pressures, take samples. Sometimes modified **lead** and **tail slurries** are used
11	Drop top cement plug	
12	Pump spacer behind	To separate drilling fluid from cement
13	Displace cement until top plug bumps	Confirm that cement is accurately placed
14	Pressure test casing	Confirm casing integrity

(*Continued*)

(Continued)

Operation	Comments
15 Release pump pressure at surface	Confirm that float shoe/collar is holding
16 Allow cement to go hard	Can take 6–48 h
17 Pressure test BOPs	While waiting on cement
18 Run **Cement Bond Log (CBL)** if required	To confirm quality of cementation bond with formation and casing

The casing string comprises some components that are specifically designed to support cementation operations or to improve the quality of the cement seal. At the bottom of the casing string, float equipment is installed to

- prevent back-flow when cement is pumped in place;
- provide a landing shoulder for wiper plugs;
- provide a guide for casing being run in-hole;
- enable pressure-test of casing after displacement of cement.

The **float shoe** is the lowermost component and serves to guide the casing along the wellbore as it is run into the hole. The **float collar** is located two casing joints (about 20 m) above the float shoe — the length of casing between the two is called the **shoe track**. Both devices are essentially one-way valves that can be pumped through but which prevent back-flow into the casing. Both are built of **drillable** components — typically cement, plastic or aluminium. Float equipment prevents the casing filling from below as it is run, requiring regular topping-up at the drill floor (see Figure 1.91). A variation of float equipment allows **auto-filling**; the one-way feature is reinstated by dropping a ball, applying pressure and shearing out part of the component.

From the sequence of operations listed above, it can be seen that rubber plugs are dropped into the casing ahead of and behind the

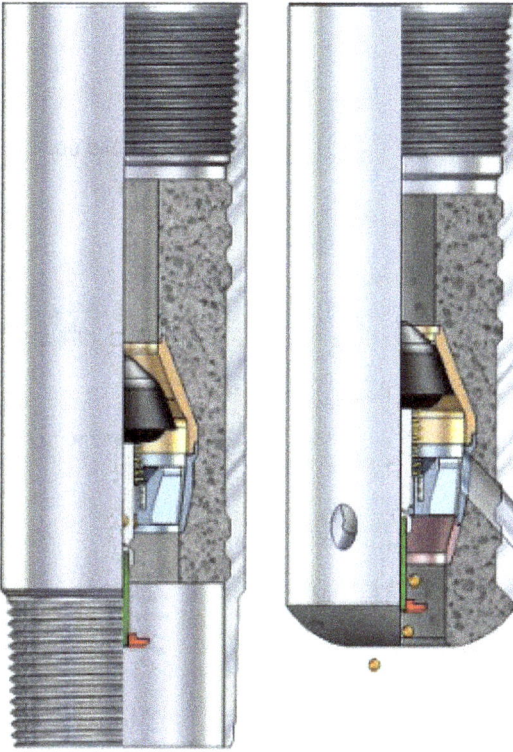

Figure 1.91. Float collar (left) and shoe.

cement slurry. The bottom plug is dropped first and precedes the cement, moves down the casing as cement is pumped and lands in the float collar near the bottom of the well. The pump pressure is then increased slightly to break the rupture disk in the top of the bottom plug after which cement is pumped into the shoe track and then into the casing/hole annulus. Once the slurry has been pumped, the top plug is dropped, the slurry is displaced using drilling fluid and eventually the top plug lands on top of the bottom plug at the float collar. Both plugs are made of rubber, colour-coded to differentiate them, with vanes to wipe cement off the inner casing bore as they are pumped through. Both are drillable, and sometimes feature *anti-rotation* features to prevent them from spinning in the casing when being drilled (see Figure 1.92).

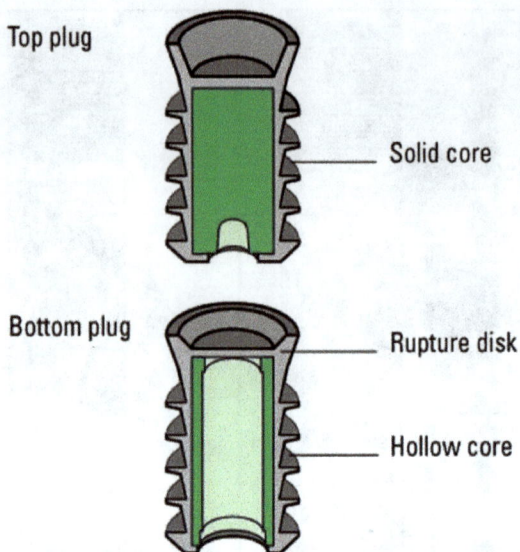

Figure 1.92. Top and bottom cement plugs.

Temporary equipment — called a *cement head* — is connected to the top of the casing to allow circulation of fluids and dropping of plugs (see Figure 1.93). This comprises a cylindrical plug holder with manifolds and valves, and a releasing pin to allow the plugs to be pumped into the casing. On offshore operations, these can be remote controlled. Operationally releasing, pumping and landing of plugs can be unreliable, so during the operation theoretical volumes of fluid pumped are compared with actual volumes to confirm that the operation is proceeding to plan.

1.16.3. *Centralisation*

Fundamental to good cementation is that the casing sits concentric in the hole and/or previous casings. If the casing rests on one side of the hole, fluid velocity during cementation on the narrow side will be much lower and achieving a good bond there will be very difficult (see Figure 1.94). In highly inclined or horizontal wells, the slide-loading on the casing can be considerable. Casing is centred in the wellbore using various types of centralisers. Centralisers

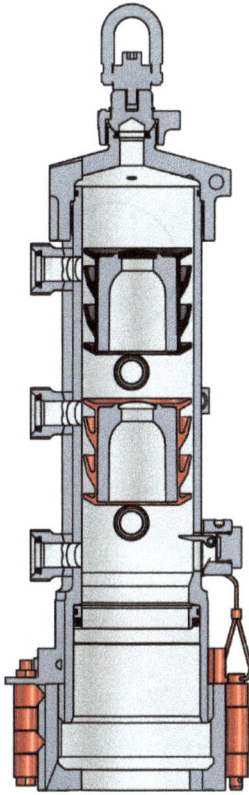

Figure 1.93. Cement head.

are positioned on the outside of the casing at regular intervals as it is run, the vanes of which can be solid (used inside a previous casing) or flexible (used to accommodate irregularities of the open hole). See the flexible centraliser in Figure 1.95. Computer programmes are used to optimise the number and placement of centralisers — too many and the flow of cement in the annulus is unnecessarily impeded, too few and the casing will not be sufficiently centralised.

1.16.4. *Scratchers and Metal Petal Baskets*

In addition to centralisers, various other devices can be run on the casing string. Scratchers are designed to remove the filter cake on the

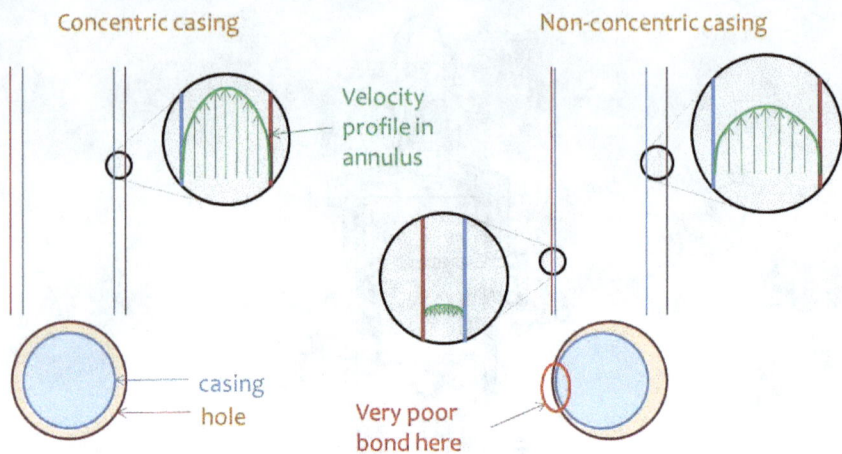

Figure 1.94. Velocity profiles in concentric and non-concentric strings.

Figure 1.95. Flexible centraliser.

wellbore face during the running of the casing to improve the cement bond with the borehole wall. Metal petal baskets work as a one-way valve to prevent the cement sloughing down the annulus once it has been pumped into position.

1.16.5. *Pump Rates Rotation and Reciprocation*

During cementation, there are a number of ways to improve the placement of cement in the annulus. High pump rates are preferred (unless this leads to losses into the formation) to achieve turbulent flow in the annulus and to avoid channelling whereby not all the mud is removed from the annulus. Bonding with the casing is thought to be improved by reciprocating the casing a few feet up and down and by rotating the casing, though sometimes this movement is not possible for practical reasons.

1.16.6. *Top of Cement*

A key aspect of cementation is achieving Top of Cement (TOC) in the annulus where it was planned. Sometimes, TOC is designed to be at surface, sometimes just inside the previous casing show, sometimes a fixed height above a hydrocarbon-bearing zone. Space precludes a detailed discussion here on the basis of these well design choices, but achievement of the planned TOC is important. TOC may be deeper than planned due to the open hole being larger than assumed or losses to the formation during the cementing operation, shallower than planned if the hole was smaller than assumed or if there was cement channelling — i.e. not completely filling the annulus. The exact size of the open hole is often confirmed by running a caliper log in the open hole prior to running casing. Accurate cement volume calculations are important to achieving a good cementation.

1.16.7. *Cementation Equipment*

Special equipment is used on the rigsite for mixing and pumping cement. On land operations, this is typically brought in specifically for the operation on a truck; offshore it is part of the permanent rig equipment (see Figure 1.96).

Dry cement powder is stored either in a bulk storage silo from which cement is aerated or in individual sacks that are opened during the operations. Mixwater is pumped by the cement pump though a venturi orifice where the dry cement is added via a hopper and the slurry created. Thereafter, the slurry is pumped by the cement pump

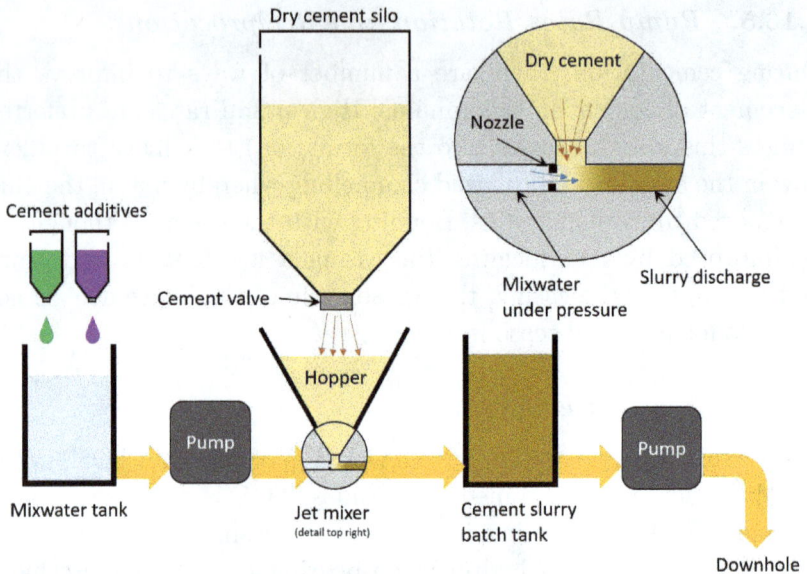

Figure 1.96. Cementing equipment.

to the cement head and into the casing. The mixwater is prepared from either sea- or fresh-water and additives are used to adjust the setting time of the cement, reducing the friction and the amount of fluid loss during the setting process. The displacement fluid — usually drilling fluid — may be pumped either by cement pump or the main rig pumps. Samples are taken of mixwater and slurry — the latter are placed into a temperature-controlled environment to replicate downhole temperature.

The process described above applies to routine casing cementation.

1.16.8. *Liner Cementation*

A liner is hung-off a previous casing string and is run and set into position on drill pipe. In this case, the plug(s) are located in the liner hanger rather than at surface. Both plugs are hollow, with different inner diameters. Just before the cement slurry is pumped, a ball is dropped and travels through the drill pipe and upper plug, and lands in the lower plug. An increase in pump pressure shears the plug from the liner hanger and it travels down the liner to the liner float collar in a similar way to the casing example above. The top plug is released in a similar way, with a larger diameter ball that lands off in the top plug.

1.16.9. *Stage Cementation*

A stage cementation is one in which the cement is placed in the annulus in stages — i.e. the lower part of the casing is cemented first, then an upper section. This requires the use of a *stage collar* in the casing string, which is a ported sleeve activated by a dropped dart and closed with a plug. Staged cementations are used when there is a risk of losses during the cement placement — i.e. cement seeps into the formation rather than circulating up the annulus between hole and casing.

1.16.10. *Subsea Cementation*

In a subsea well, the casing hanger lands off in the wellhead which is located at the seabed rather than at surface. Subsea casings can be run with casing above the hanger, which is then disconnected from the hanger after the cementation. Alternatively, a subsea cement head can be used in which the bottom and top plugs are retained within the wellhead running tool and released by sending a sequence of darts or balls down the drill string from the rig.

1.16.11. *Surface Casing Cementation*

For very large casings (30″ and 20″), it can be impractical to pump and displace the cement down the entire (large) internal bore of casing. The low linear velocities achieved can contaminate the cement. Often a **stinger cementation** is used whereby the casing is run and landed and thereafter a drill string is stabbed into the float collar. The complete cementation sequence then takes place down the drill string rather than the casing. Typically, no plugs are used.

1.16.12. *Top-Up Cementation*

For surface casings and conductors, top-up cementations can be achieved in which cement is pumped into the annulus from above via hoses. This is typically required to achieve structural stability of the wellhead rather than a seal between casing and wellbore.

1.16.13.   *Setting a Balanced Cement Plug*

There are times when a cement plug needs to be placed in the well, either in open- or in cased-hole. Gelled mud is often positioned in the well first to prevent the cement sloughing down the hole. The technique involves running a **cement stinger**, which is often a short section of tubing, on drill pipe. Cement is then pumped into position so that the top of the cement is at the same depth in the annulus as in the stinger. The stinger is then pulled out to the top of cement, and the well reverse-circulated to clean-out excessive cement. The cement is then allowed to set. Confirmation of a good plug is made by setting weight down on it, and/or by pressure testing or inflow testing it to ensure a pressure-tight seal.

Cement plugs are often used during the **abandonment** of a well, to isolate a water-producing zone, to make a well safe for suspension or to start a side-track.

A variation of the plug setting procedure is used for a **cement squeeze**. There can be occasions such as when a casing cementation has not been successful, when cement must be squeezed into the annulus or into the formation. A technique similar to balanced cement placement is used. Holes (perforations) are made in the casing at the squeeze point. The cement is placed opposite the squeeze point and the stinger retracted, but before the cement hardens, pressure is applied to squeeze some of the cement into the annulus and/or the formation as required. A pressure higher than formation breakdown pressure may be needed. The excessive cement in the wellbore can be flushed or drilled away later.

1.17.   Directional Drilling

One of the key requirements for a well is that from a surface location it reaches the required position in the reservoir. We drill wells directionally for the following reasons:

- access one or more specific reservoir sections/zones;
- access difficult reservoir locations:
  - under residential area/natural park;

- under a lake/near shore;
- avoiding difficult to drill or troublesome formation, e.g. mobile salt;
- avoiding high pressure;

- allow multiple wellheads from one surface location:
  - offshore platform/jacket offshore, up to 60 wells (tight spacing);
  - pad drilling land wells, reducing environmental impact;
  - sub-sea template;

- relief well drilling — to intersect an existing well;
- highly deviated/horizontal wells to increase exposed section length through the reservoir to allow higher production.

Setting surface location and downhole targets is the first step of planning a well.

The subsurface location is determined from seismic data, previous wells, the reservoir model and petrophysical logs. Criteria for selecting targets may include:

- optimal reservoir drainage;
- distance from faults;
- lateral extension from a surface location;
- position relative to other targets;
- high-pressured areas (such as below the target).

The surface location is determined by:

- natural surface features such as rivers, hills/valleys;
- man-made considerations like distance from roads, housing, parks, beaches;
- licence boundaries;
- environmental impact;
- distance to processing facilities;
- offshore: water depth, common well locations like platforms, seabed conditions.

There may also be obstacles between surface and target(s) that need to be considered, such as:

- pore pressures and rock strength (setting depths of casings);
- wellbore instability due to *in situ* stresses;

- shallow gas hazards;
- presence of faults;
- other wells;
- limitations in drilling the well (e.g. rate of turning);
- optimisation of the well design to minimise cost, maximise production, reliability and retrieval of downhole information.

This wide range of considerations points to the need for careful planning.

A grid system is used to determine required coordinates — typically denoted as Nothing and Easting coordinates on a **Universal Transversal Mercator (UTM)** grid — a typical reference would be 17N 630084 4833438 — i.e. on the 17 grid in the northern hemisphere, northing 630,084 m, easting 4,833,438 m.

As a datum, depths are often referenced to **Mean Sea Level (MSL)** over a field. Everyday depths during a drilling operation are referenced to **Drill Floor Elevation (DFE), Rotary Table (RT)** or to the **Rotary Kelly Bushing (RKB).** Other reference depths are used during the production phase such as the top of the wellhead flange.

Depth and location references are also of crucial importance — there are examples of wells being drilled in the wrong place or to the wrong depth due to errors in applying these systems.

In a vertical well, the target(s) are vertically below the surface location and the well is drilled in a straight line between these points. Wells with offset targets typically include several of the following as shown in Figure 1.97:

- **Kick-off Point (KOP)**
- **Build-up section**
- **End Of Build (EOB)**
- **Tangent section**
- **Drop off section**

A well is typically drilled with a vertical section and surface casings set above the KOP. A different drilling assembly is used to build inclination in the build-up section. Once the required inclination is

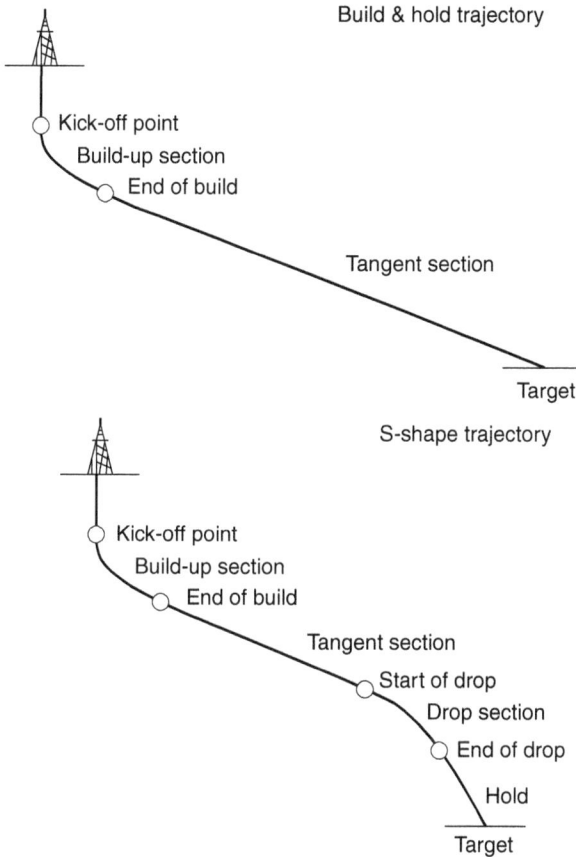

Figure 1.97. Well trajectories.

achieved at EOB, another casing may be run. Once again a different assembly is run in the tangent section, where the inclination of the well is held constant. This may continue to the target. Sometimes, an S-shaped well is drilled in which the well is allowed to drop off and held as required until the target is reached. This may be done to improve wireline logging over the reservoir section. Throughout these various sections, the wellbore may also turn left or right. Targets are normally specified as a three-dimensional shape with central northing, easting and depth coordinates and tolerances on each of these coordinates. Shapes such as spheres or cubes are common.

There is a trade-off on target sizes — optimal reservoir drainage versus cost and time to drill which needs to be resolved between the subsurface and well engineering staff during the well planning phase; this may be revisited during drilling operations if a target is liable to be missed.

Used to combat blowouts, **relief wells** are a special case of directional drilling when it is required for one well (the relief well) to intersect another (the blowing well) in order to pump kill-weight drilling fluid into the blowing well and stop the flow. These wells are very dependent on detailed planning and the use of **homing tools** to detect and find the blowing well.

A directional drilling well plan is shown in Figure 1.98. This comprises two sections. The vertical section view shows the required well trajectory projected onto a vertical plane as defined by target and surface location. In this case, the casing setting depths are also shown. In the plan view, it can be seen that the well will be drilled in a roughly North West direction, to the targets that are circular in plan view.

These plots are generated from well design software that computes the various well positions from modelling software using a technique called the *minimum curvature* method. In addition to plots like these, data are provided in tables.

The well trajectory design software is combined with the local knowledge of the well engineer to develop a "drillable" wellbore, making the necessary compromises between cost and future well productivity. Here are some examples for consideration:

1. A maximum **dogleg severity** (a measure of how tightly the wellbore turns) may be specified to ensure that casing can be run in the well once drilled.
2. The well inclination may be reduced in the reservoir to avoid problems when logging.
3. Changes in inclination and azimuth may be combined so that rotary steerable systems can be efficiently deployed.
4. Casing may be set in tangent sections to avoid problems when drilling out the casing "rathole".

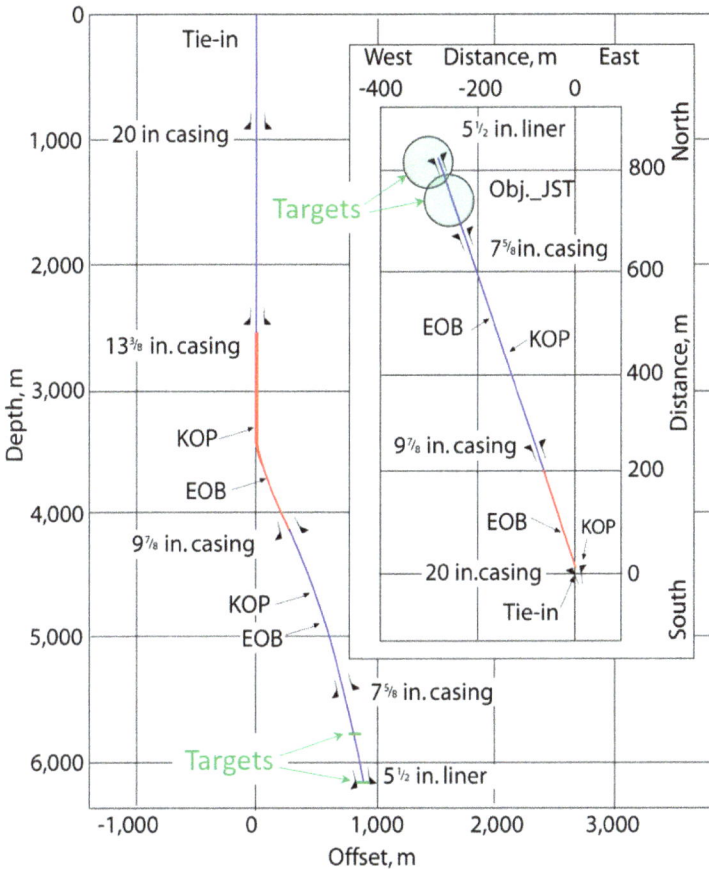

Figure 1.98. Deviated drilling plan and section.

As the well will is drilled, the trajectory views are updated with actual positional data for the well — the process is covered in the next section.

The *inclination* of the wellbore (i.e. in the vertical plane) is described as degrees from the vertical. The direction of the wellbore (i.e. in the horizontal plane) is described as degrees from **Grid North (GN)** or **True North (TN)**. Note that this is not usually the same as **Magnetic North (MN)**. An example of the difference is shown in Figure 1.99. These variances are location-specific, and in the case of magnetic north also time-specific.

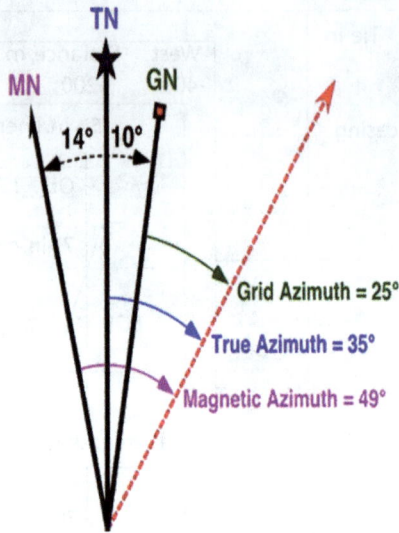

Figure 1.99. Example radial offsets — true/grid/magnetic North.

1.18. Directional Surveying

Knowing the location of the wellbore and its direction is important for the following:

- managing the drilling of the well to the target as per the design plan;
- ensuring that data from the well is correctly mapped on geological and reservoir subsurface models;
- avoiding collisions with existing wells;
- to provide a target should a relief well be required;
- avoiding transgressing adjacent licence areas or properties;
- provision of positional data is frequently a regulatory requirement.

The following summarises the various techniques and tools used to survey a wellbore:

The **surface location** is confirmed using a variety of conventional techniques such as GPS and geometric surveying from other known reference positions.

The **Totco survey** is the simplest technique still in use. It comprises a cylindrical tool about 3 cm in diameter and 80 cm long that contains a pendulum and indicator card. Drilling is halted for a few minutes and the tool is run on wireline or dropped down the drill string and comes to rest in a ring (called a Totco ring) that centralises the tool inside the drill string. After a fixed period of time (or nowadays once the tool detects no motion), the indicator card is marked by the pendulum which provides direct information on inclination. The tool is then retrieved using wireline (or in the BHA if recovering it) and the card removed to extract the data.

The **Magnetic Single Shot (MSS)** tool is similar in principle to the Totco and is run in a similar way. Historically, it included a compass and a camera in addition to the pendulum. The camera takes a picture of where the pendulum points to on the compass face. This provides direct information on inclination and azimuth.

Magnetic Multi Shot (MMS) tools comprise a similar system to the MSS except that multiple surveys can be taken. It is run on wireline or dropped just before retrieving the drill string and takes surveys as the string is recovered.

Nowadays, solid state tools are more common, which are far more reliable and accurate and use orthogonal gravimeters and magnetometers. Data are recorded in electronic memory and/or transmitted back to surface on wireline if so run.

All magnetic survey tools require a non-magnetic environment. The survey tools and drill collars in the BHA must be non-magnetic to avoid misreadings from the magnetic compass, and for this reason surveys can't be taken inside casing or very close to other wells. Company standards will specify how often a survey is required, but typically this might be every 500 to 1,000 ft. All azimuth data from magnetic surveys must be adjusted to grid north from magnetic north.

1.18.1. *Gyro Survey Tools*

Gyro-based surveying tools have historically used physical gyroscopes to measure the direction of the wellbore, and were run on wireline due to their fragile nature. The advantage of the gyro tool is greater accuracy than the magnetic survey, and because there is no interference from irregular magnetic fields it can be run in casing or in drill pipe. Because it uses a different physical principle, gyro tools offer independent confirmation of previously-acquired magnetic survey data.

Once again, modern tools use solid-state gyros and electronic memory.

1.18.2.   *Inertial Navigation Tools*

These tools are the most accurate and were originally developed from defence missile navigation systems. A gyro-stabilised platform is used on which three orthogonal accelerometers are located. The signal from the accelerometers is integrated twice with time to provide x, y and z coordinates directly, reference the hole at surface. These tools are large and expensive.

The depth reference for all surveys bar the inertial navigation tools comes from a wireline depth or drill pipe depth. A significant amount of QA/QC is required on all directional logging tools, and calibration of the tools is required at the workshop and at the rigsite.

1.18.3.   *Steering Tools*

Before the development of MWD tools (see below), directional drilling required the running of a surveying tool on electric wireline to determine the orientation of the bent-sub above the mud-motor (see below). This was achieved using a magnetic or gyro-based steering tool landed in an oriented shoe in the non-rotating drill string above a mud-motor. It provided real-time readout on the drill floor, showing tool-face orientation relative to the high side of the well, in addition to conventional survey data. These tools are still used occasionally for setting oriented packers and for oriented coring operations.

1.18.4. *MWD Tools*

MWD tools are frequently used in modern-day drilling operations. The MWD tool comprises a non-magnetic drill collar in the BHA located just above the bit. The wall of the collar contains solid-state magnetic sensors, memory and batteries. The directional data — including tool orientation — is retained in memory. A valve is opened and closed electrically based on the data, and this creates pressure pulses in the drilling fluid that can be read at surface. Hence, real-time directional data is provided at surface without wireline. This allows virtually continuous steering of the wellbore.

1.18.5. *Homing Tools*

Homing tools are used in the event of needing to drill one well into another, for example to intersect a blowing well. The most common technique is to detect abnormalities in the local magnetic field created by the blowing well (provided of course that it contains ferrous material). The intersecting wellbore is then steered in the direction required, and resurveyed regularly to confirm progress towards the target.

1.18.6. *Survey Uncertainty*

Every survey carries a degree of uncertainty. Generally, the costlier the survey tool, the more accurate the result. Each tool has an error model that is based on the physical characteristics of the device, and the environment in which it is run. Suffice to say here that each survey point can be considered to be at the centre or an ellipsoid of uncertainty (see Figure 1.100). There is as 99.9% certainty that the wellbore actually lies within this ellipsoid.

This uncertainty is important. Although from a survey it may appear that a target has been hit, if the ellipsoid extends beyond the target there is a possibility that the wellbore actually lies outside the target. This is managed in reality by taking the **geological target** and reducing it in size depending on the uncertainly of the surveying programme to provide a **drilling target** that is smaller. If the centre

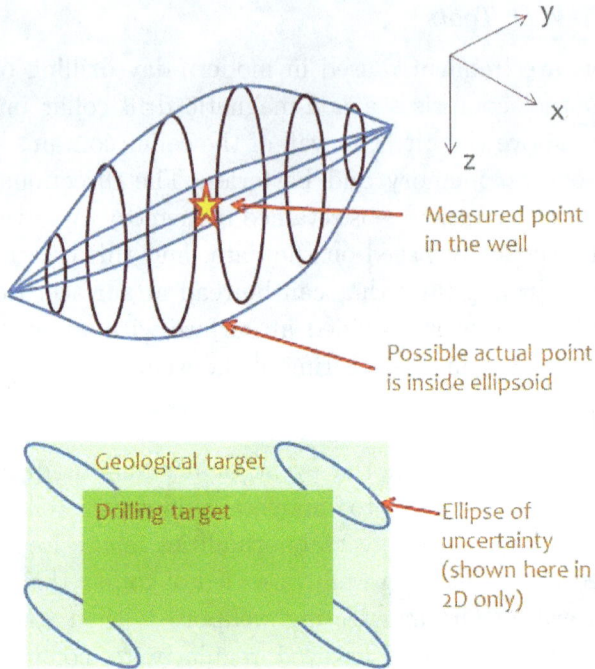

Figure 1.100. Ellipsoids of uncertainty.

of the wellbore passes through the drilling target, there is certainty that the geological target was hit (see Figure 1.100).

Where many wells exist from a platform or multi-well pad on land, it is essential when drilling a new well that it does not intersect with another. Sometimes, wells have to be directionally drilled through a "spaghetti mesh" of existing wells. The position of existing wells and their uncertainty must not overlap with the position of the drilling well and its uncertainty; otherwise, there is a risk of collision. The most accurate tools are used to survey wells in these cases. Operators and regulators set standards that determine how close the ellipsoids of uncertainty can get before drilling must stop. In reality, nearby wells are often closed-in below the point of potential intersection if there is any risk of intersection.

1.18.7. *Plotting and Recording Survey Data*

During drilling activities, survey data is collected and a smooth wellbore trajectory calculated. The **minimum curvature** method is used. The actual trajectory is plotted as soon as possible against the planned trajectory in both plan view and section view. From this, decisions can be made to steer the well in a different direction if required.

Once the wellbore section is complete, an independent (e.g. gyro) survey is carried out to confirm the wellbore position, and being more accurate this survey is used as the *definitive survey* along with its uncertainty and is formally recorded as the location of the well in company and government databases.

1.19. Directional Drilling Systems

Modern drilling operations have been transformed by directional drilling capabilities.

Earliest directional drilling techniques involved "badgering" — using a jet of drilling fluid from the side of the drilling assembly to push it in a certain direction. This is similar in principle to the steering tools we have today — covered below.

In fact, any drilling BHA rotated in a hole will have a tendency to (i) drop or build inclination and (ii) turn left or right. The former, in particular, can be predicated with good accuracy. As illustrated in Figure 1.101, placement of stabilisers along the BHA, if the well is already inclined, will act as a fulcrum and cause the bit to point up or down.

A **packed BHA** is one with several full-gauge (hole-sized) stabilisers, forcing the bit to point along the wellbore. A **pendulum BHA** has no near-bit stabiliser and in an inclined wellbore the bit has a tendency to "point down" and make the hole more vertical. However this effect is dependent on weight on bit — increase the WOB and the tendency will be to drop less or even build. A **fulcrum BHA** — with a full gauge near-bit stabiliser (same outer diameter as the bit) — will cause the bit to "point up" and the wellbore inclination will be increased.

These principles are used when drilling conventionally — the lowest cost approach and one frequently used when only fine changes of inclination and no changes in azimuth are required.

1.19.1.   *Bent-Subs and Mud-Motors*

For more significant changes in wellbore direction, the industry has developed the mud-motor/bent-sub combination (see Figure 1.102). As can be seen from these examples, the bit is driven directly by a mud-motor and connected back to the BHA and drill string to surface via a bent-sub. The mud-motor is inside the near-bit stabiliser in these examples. Various types of bent-sub are available, frequently allowing adjustment at surface and even downhole. Bend angles of up to 3° are normal.

1.19.2.   *Mud-Motors and Turbines*

A mud-motor comprises a smooth lobed shaft with a twisted profile that sits into a rubber stator with one more lobe than the rotor (see Figure 1.103). Drilling fluid passing through the BHA flows in the spaces between stator and rotor, causing the rotor to turn and this is connected to the bit by a shaft that runs inside the near-bit stabiliser. A dump-sub directly above the motor is a sleeve that can be operated by dropping a ball prior to pulling out of hole to allow drilling fluid in the drill string to drain easily as it is recovered. A drilling turbine performs a similar function to the mud motor, but works instead using a series of fixed stators and rotating rotors as in any other type of turbine. Mud motors generate high torque at relatively low RPM; turbines develop low torque at high RPM, and hence are suitable for drilling with diamond bits for example. In both types, *reactive torque* is transmitted up the drill string to surface (see Figure 1.104).

About 30 years ago, the most advanced drilling using motors with bent-subs involved using a magnetic steering tool with surface readout that was set in an oriented profile above the bent-sub. This was used to orient the bent-sub by turning at surface, whereupon

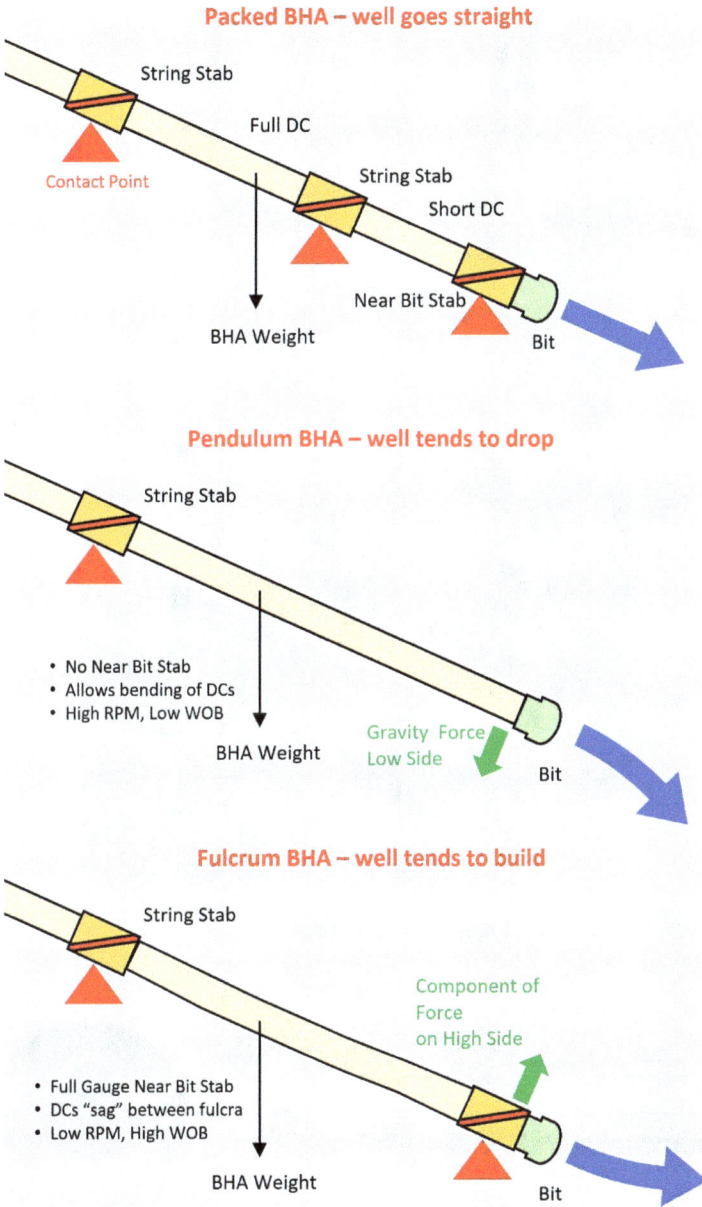

Packed BHA – well goes straight

String Stab

Full DC

Contact Point

String Stab

Short DC

Near Bit Stab

BHA Weight

Bit

Pendulum BHA – well tends to drop

String Stab

- No Near Bit Stab
- Allows bending of DCs
- High RPM, Low WOB

BHA Weight

Gravity Force
Low Side

Bit

Fulcrum BHA – well tends to build

String Stab

Component of
Force
on High Side

- Full Gauge Near Bit Stab
- DCs "sag" between fulcra
- Low RPM, High WOB

BHA Weight

Bit

Figure 1.101. Principles of BHA deflection.

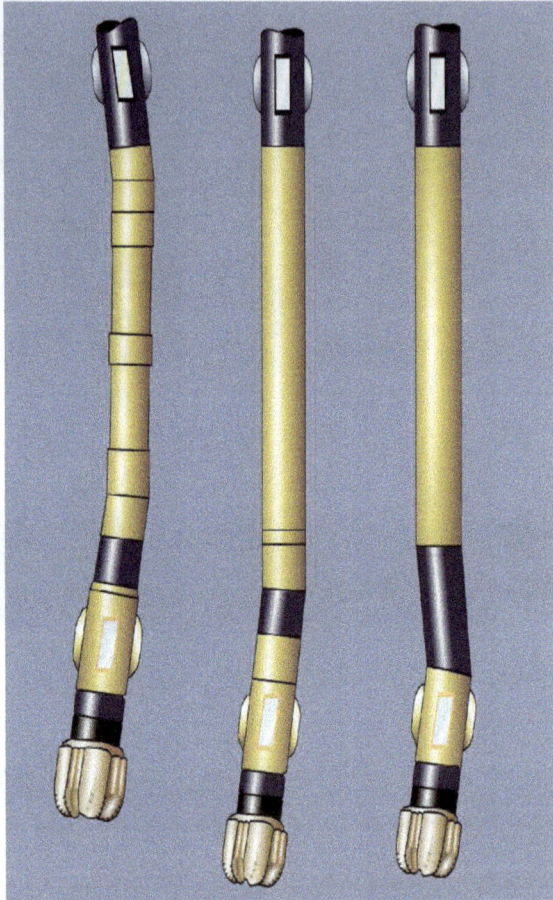

Figure 1.102. Bent sub types.

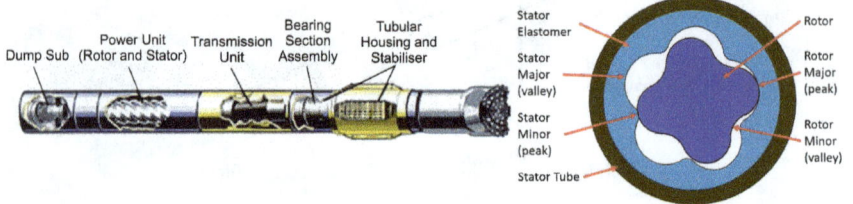

Figure 1.103. Mud motor design.

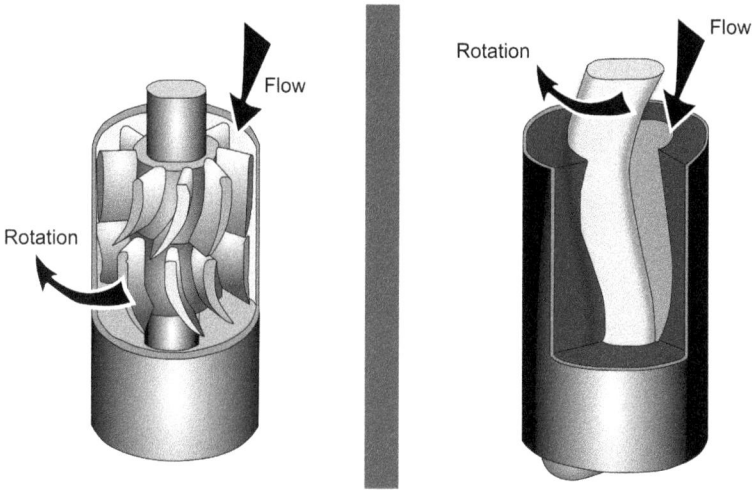

Figure 1.104. Mud turbine and motor details.

drilling with the mud-motor could commence. Directional surveys would be taken frequently and the directional driller would make changes to the orientation of the bent-sub depending on progress. He would need to account for the reactive torque response from the motor — requiring quite a bit of judgement. In this mode, there would be no rotation of the drill string at surface (other than to orient the sub). The same BHA could also be used in "sliding mode" in which the whole assembly would be turned from surface and the bit RPM would then be the sum of surface RPM and motor RPM, which speeds up ROP.

1.19.3. *MWD and LWD Tools*

We have touched on MWD tools for surveying above and will now cover them in more detail. MWD tools are equipped with the following sensors:

- Basic trajectory parameters:
  ○ inclination;
  ○ direction;
  ○ tool-face.

- Additional drilling sensors may include:

  o pressure (inside BHA and annulus);
  o temperature;
  o weight on bit;
  o bit RPM;
  o bit torque.

- The LWD tool may include petrophysics sensors:

  o gamma-ray;
  o resistivity;
  o sonic;
  o density;
  o other petrophysical evaluation tools.

The tool is designed to store this data in memory and also to transmit it back to surface in real time.

The tool comprises the following sections:

The **power supply** comprises as set of lithium batteries rated for the downhole temperatures expected. A small **turbine generator** positioned in the mud flow may also be provided to generate power from the fluid flow downhole and to recharge the batteries. The **directional sensor** section usually comprises three accelerometers and three magnetometers. The **logging section** comprises the measurement tools to measure the petrophysical data. These two sections and other electronics are located in the wall of the MWD/LWD tool collar, which is machined from solid non-magnetic steel. Three alternative **data transmission** systems are in use — **mud pulse** (data rate 40 bps), **acoustic** (50 bps), **ElectroMagnetic** (**EM**, 10 bps), and **wired pipe** (50k – 1 M bps).

The mud pulse system uses a valve in the mud flow or bypassing it that creates a positive or negative mud pulse that is detected at surface as a change in circulating pressure of 10–50 psi. For steering operations, this data is continuous, for survey or logging information the data is compressed and encoded for efficient transmission as well as being kept in tool memory as a backup. There is no practical depth limitation of this technology, but it is dependent on good quality drilling fluid.

The EM system transmits data via magnetic pulse or electrical current through formation to surface antennae. Data can be transmitted at any time, but there are limitations regarding depth and the type of formation drilled. However, EM can be used if the drilling fluid is air or under-balanced drilling operations are being carried out.

Wired pipe is a new technology that requires the use of drill string components equipped with conductors to transmit data to and from the tool. Such components are expensive but are demonstrating improving reliability. The system allows much higher quality real-time data to be acquired while drilling.

1.19.4. *Rotary Steerable Systems*

The current state-of-the-art in directional drilling is the Rotary Steerable System (RSS) — such as that shown in Figure 1.105. The upper components of the tool are similar to the MWD and LWD tools already described. The tool can be used in combination with a motor or in rotary mode (rotation from surface). At the top left

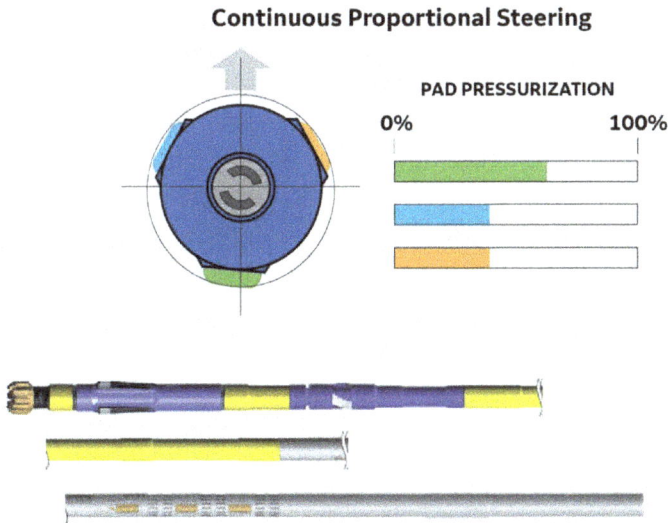

Figure 1.105. Rotary steerable assembly.

is shown a section through the tool; the non-rotating sleeve (shown in blue) is held off the borehole wall using three pads that can be pressurised to push the bottom part of the BHA in any desired direction relative to the well, hence changing the direction of the hole being drilled. Control of the pistons is managed by an in-tool computer system based on instructions typically communicated to the tool based on stops and starts of the mud-pumps. Both "push the bit" (as described here) systems and "point the bit" systems are available.

In summary, the RSS/LWD combination provides the following advantages:

- allows continuous pipe movement for faster drilling and less likelihood of getting stuck;
- supports better hole cleaning as pumping is continuous;
- reduces/eliminates trips for assembly changes;
- permits more complex well paths to be drilled;
- allows well track to be kept close to plan;
- enables geosteering;
- makes hitting smaller targets easier;
- permits extended reach drilling;
- can be run with a range of petrophysical tools.

Whipstocks (see Figure 1.106) are used when the well needs to be sidetracked. Sidetracking is the drilling of the well out from an existing casing and into adjacent formation, e.g. to access hydrocarbons away from the original wellbore or to obtain a core through a previously drilled formation.

A whipstock has a tapered element at the top, and a set of slips like a packer below. It is run into the hole on a drill string, oriented so that the face of the tapered section faces the intended side-track direction and set in the casing at the required depth. If the well is not vertical, a side-track may take place to the "high side" or the "low side" of the hole (see Figure 1.107).

The side-track drill string assembly typically comprises a mill rather than a bit. The mill "kicks off" the whipstock taper, cuts through the wall of the casing and starts drilling a new hole alongside the existing well. The assembly is then normally pulled out and a

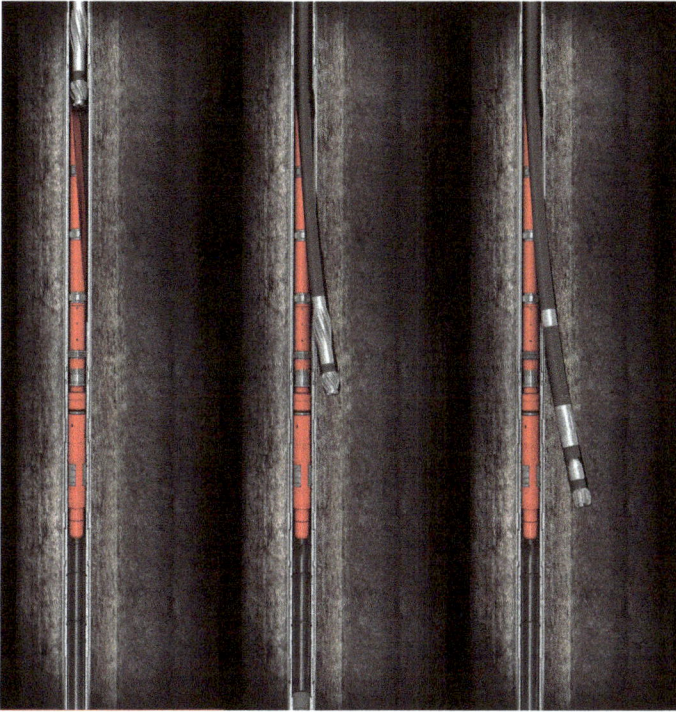

Figure 1.106. Whipstock used for side-tracking.

standard drilling assembly then run and used to drill the well which thereafter progresses conventionally.

1.20. Trajectories

Basic vertical, inclined and s-shaped wells have been included above. But modern directional drilling techniques are capable of drilling far more complex well-paths.

It is important to retain a sense of scale. The deepest wells are akin to drilling with spaghetti down from the top of a soccer stadium at one end, to halfway down the pitch.

Take a look at a typical complex well path (in this case one drilled in the Norwegian offshore sector) in Figure 1.108. The well starts vertically from a floating drilling unit, builds to a high inclination, and turns clockwise in a large arc down to reservoir depth (the formation is shown in yellow in the section). One side-track (shown

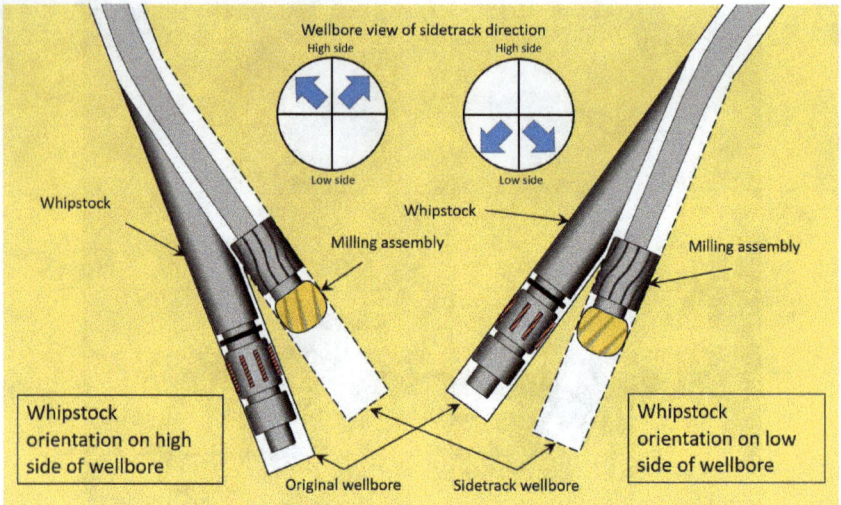

Figure 1.107. Sidetracking to high and low side.

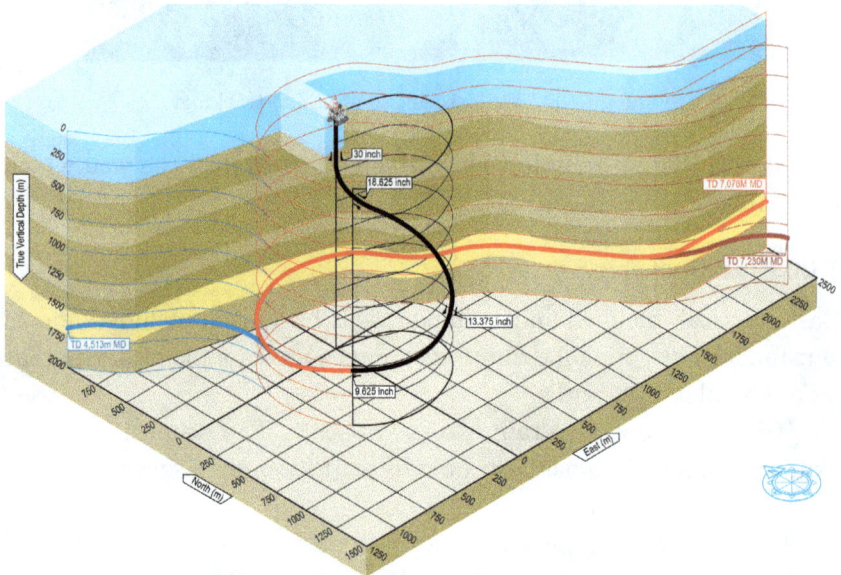

Figure 1.108. Extreme deviated drilling.

in blue) was drilled initially, probably to confirm the lateral extent and to determine the quality of the reservoir in that part of the field. Thereafter, the well was drilled horizontally in the reservoir to TD (shown in red). It is likely that this section was geo-steered to remain in the reservoir. A second side-track was drilled to a new TD perhaps to confirm the actual depth of the reservoir formation.

This type of complex wellpath is possible using modern directional drilling techniques and is designed to optimise the information gained and/or production from the subsurface. With such wells costing perhaps $100 million, it is essential that value is optimised in this way.

On 28 January 2011, the world's longest borehole was drilled at the Odoptu field, on Sakhalin Island, in Russia with a measured total depth of 12,345 m (40,502 ft) and a horizontal displacement of 11,475 m (37,648 ft).

At the other extreme, shale gas developments require wells that intersect with as many fractures in the formation as possible. Given that these fracture networks are often vertical; this requires horizontal wells drilled with close spacing. Figure 1.109 shows a typical

Shale Gas Horizontal Development

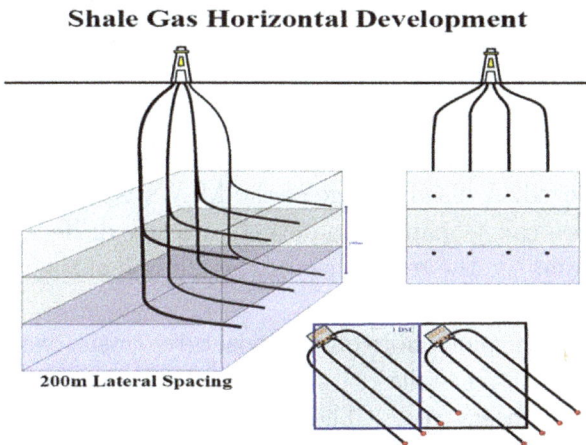

Figure 1.109. Shale gas development wells from single pads.

layout of such wells, in this case drilled from pads containing four wells. These wells are multi-laterals — i.e. there is one "mother-bore" from which side-tracks are drilled. See more on multi-lateral wells below.

Extended reach wells are used to access reserves laterally displaced from the drilling centre.

Figure 1.110 shows the Wytch farm field (originally drilled by BP), illustrating how wells are drilled from land locations south of Poole in the UK to downhole locations (shown as red dots) up to 10 km laterally offset from the surface location. In this case, expensive offshore development structures, and significant environmental impact, were avoided.

Other advanced well trajectories are shown below. The **hook well** (see Figure 1.111) demonstrates that wells can be drilled with an inclination greater than 90°. In this case, the well is drilled *upwards* to access oil trapped by inclined faults from an existing platform location offshore Brunei. Inclinations up to 167° have been achieved.

The **snake well** (see Figure 1.112) can be used to access several targets in a horizontal plane, perhaps oil in adjacent fault blocks. In this case, lateral steering is used to steer left and right; again, geosteering would typically be used.

Multi-Lateral Wells (MLWs) can be used to access different targets from one "mother-bore" (see Figure 1.113). This can be advantageous compared with individual wells because the overburden formation needs only to be drilled once. There are various classifications of multi-lateral wells which relate the architecture of the junction between the various bores — specifically to the degree of sealing achieved. A disadvantage of the MLWs is that problems in a later wellbore can jeopardise the others already drilled. Subsequent re-entry to manage the reservoir over the lifetime of the well can also be troublesome.

Figure 1.114 plots horizontal departure versus vertical depth of the most extreme wells drilled. This "ERD envelope" has been extended over recent years. The wells with the greatest departure reach to 36,000 ft (at 4,000 ft vertical depth); those of greatest vertical depth have reached 40,000 ft.

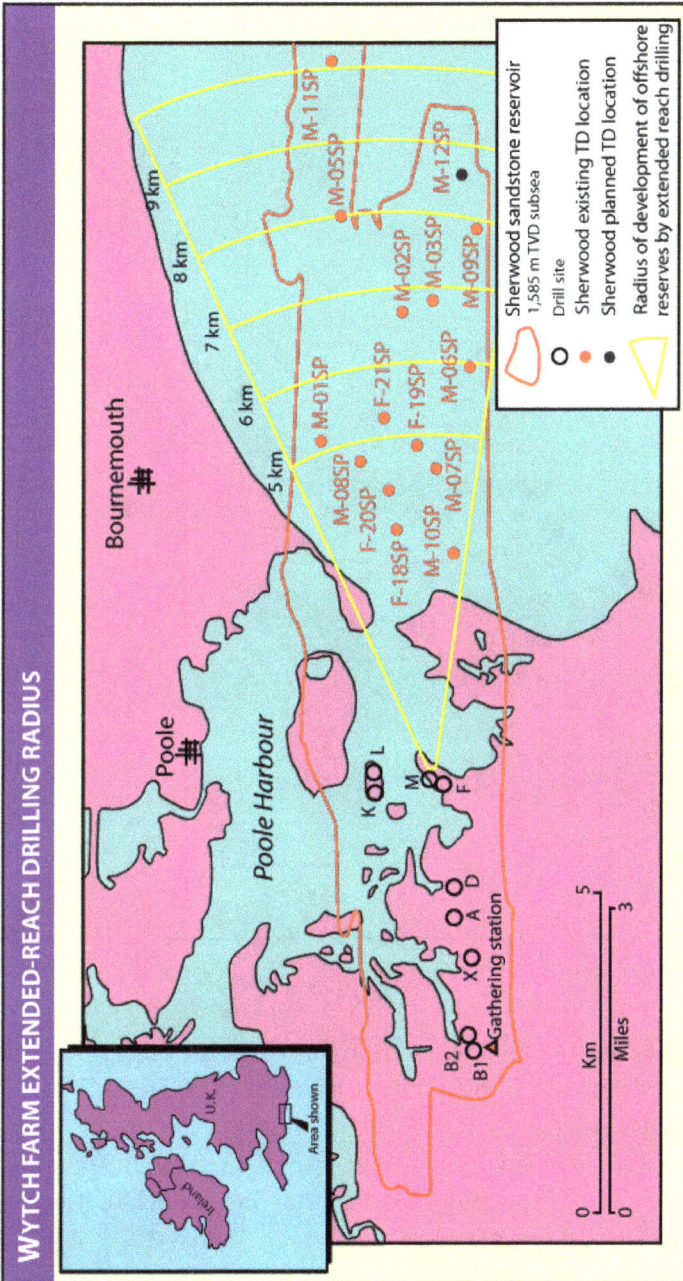

Figure 1.110. Wytch farm extended reach (ERD) wells.

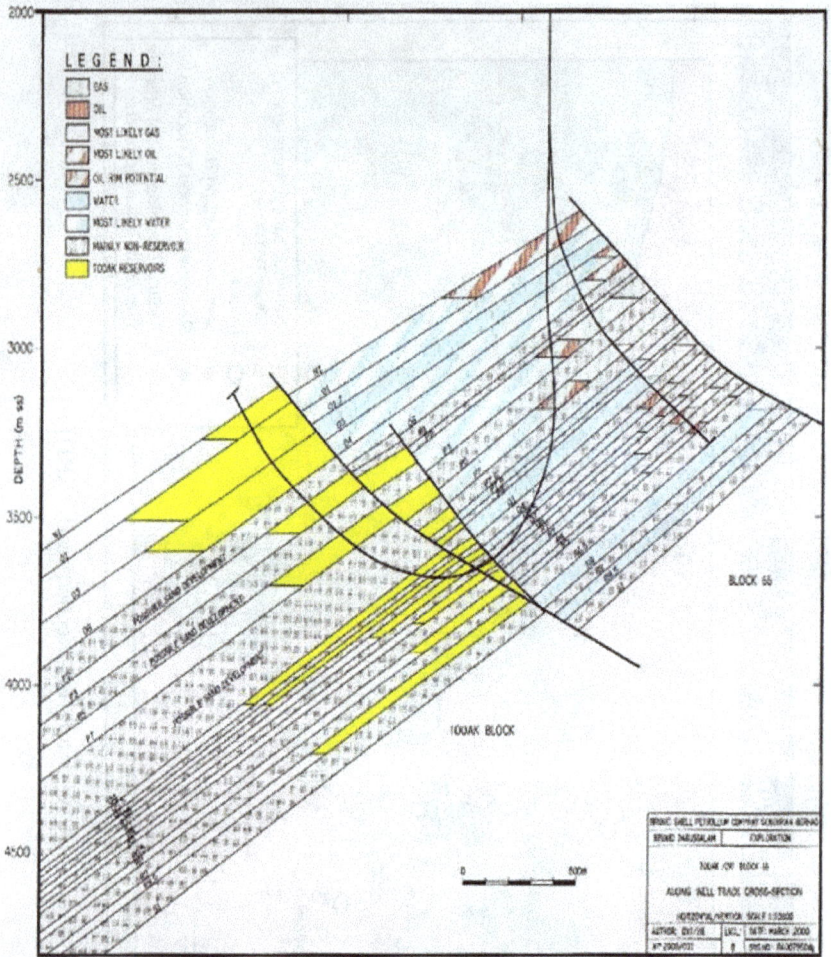

Figure 1.111. Hook well.

In some cases, parallel wells are required in the reservoir. A good example is Steam-Assisted Gravity Drainage (SAGD) wells that are used for the production of heavy oil. Steam is injected in one wellbore to heat-up the reservoir and reduce the viscosity of *in situ* oil, assisting flow to the production well that is parallel to the steam injection well (see Figure 1.115).

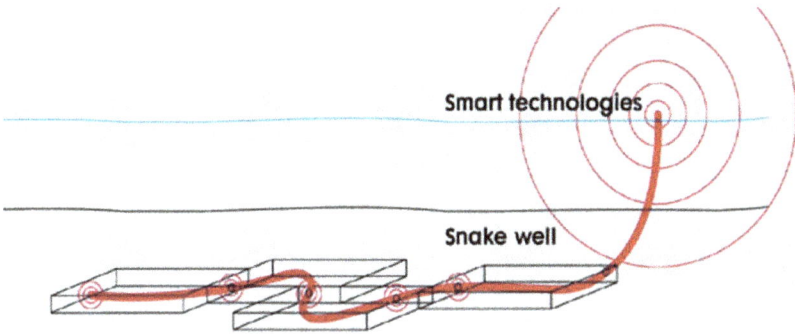

Figure 1.112. Snake well.

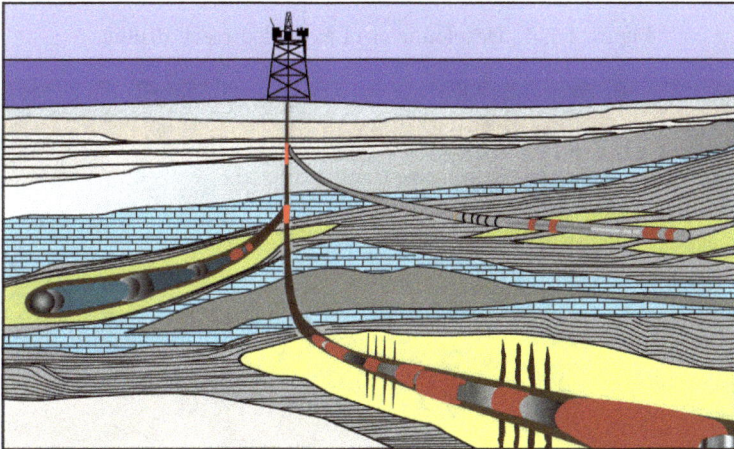

Figure 1.113. Multi-lateral well.

A logging tool that produces a strong electromagnetic field is run into the first well to provide a reference against which the second well can be drilled parallel using MWD technology.

1.21. Stuck Pipe

Most of the material in this chapter is about what happens when operations proceed smoothly and to plan. This section covers one example of problems that occur when drilling — getting the drilling assembly stuck in the hole — or more simply "stuck pipe". Given

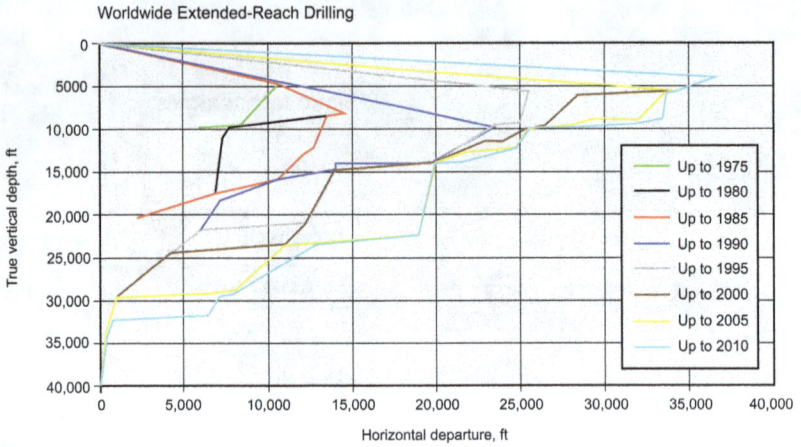

Figure 1.114. Development of extended reach drilling.

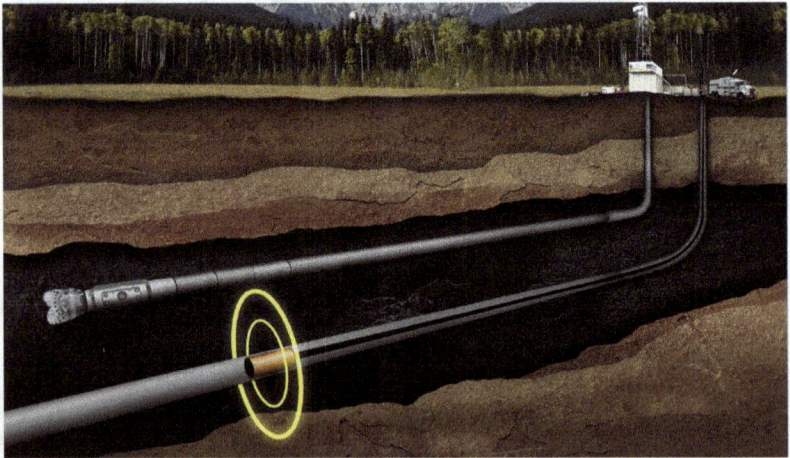

Figure 1.115. Steam-assisted gravity drainage wells.

the likely significant investment in the well at that point, and the potential for further problems if the hole is not made safe, the well cannot usually just be "re-spudded".

The following causes of "stuck pipe" will be explained in more detail below:

- differential sticking;
- poor hole cleaning;

- chemically active formations;
- drill string vibration;
- irregular borehole geometry;
- mechanical instability;
- fractured and faulted formations;
- unconsolidated zones;
- key seating;
- geo-pressured formations;
- salt formations;
- cement-related problems;
- under-gauge hole or junk.

Differential Sticking is the most common mechanism. See Figure 1.116 showing how filter cake on the side of the borehole can interfere with the BHA. Differential sticking is caused by the differential pressure between porous and permeable formation and the drilling fluid in the hole. This is greatest when drilling depleted reservoirs or when excessive mud weight is used. See Figure 1.116 showing the cross-section of a well being drilled. In normal drilling operations, mud hydrostatic pressure (P_h in Figure 1.116) exceeds formation pressure (P_f) — the pressure differential is usually designed to be 200–300 psi, but in depleted formations it can be much higher. The differential pressure in permeable zones forces mud filtrate into the permeable rock, leaving behind a wall cake. When a pipe comes into contact with the wellbore, the pipe surface in the wall cake is exposed to the lower pressure of the formation while the rest of the pipe surface is exposed to mud hydrostatic pressure. The differential force pulls the pipe firmly against the wall of the wellbore.

Thicker wall cake results in a larger area exposed to the lower pressure of the formation. Differential sticking normally occurs when the pipe is static. More cake is deposited to form a bridge, causing significant increase in effective contact area. Differentially stuck pipe continues to get more stuck as time passes. Quick action — moving or rotating the pipe — is necessary to free it.

Methods to avoid this and other types of "stuck pipe" are covered below.

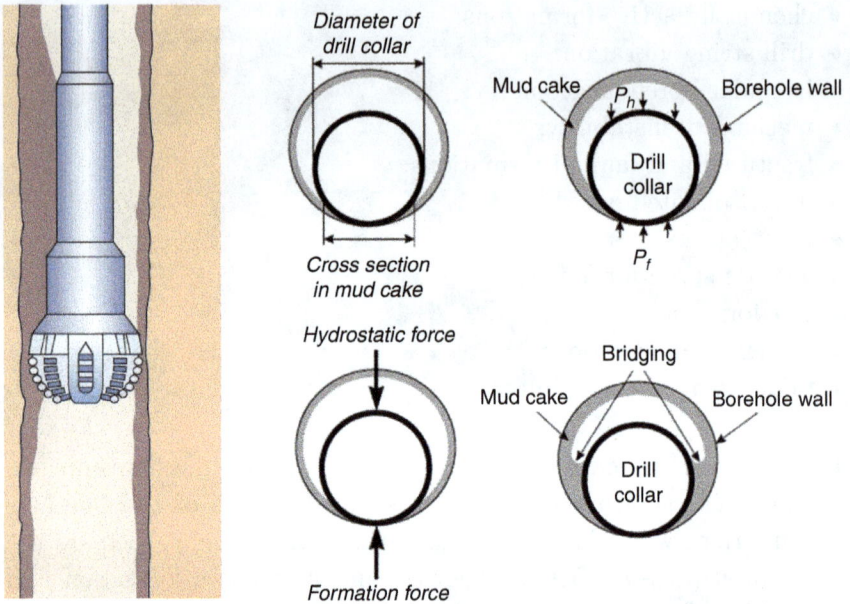

Figure 1.116. Differential sticking. Illustration copyright Oilfield Review, used with permission of Schlumberger.

Poor Hole Cleaning (see Figure 1.117) is a frequent problem due to insufficient pump rate resulting in annular velocity less than that required to carry cuttings out of the well. It is avoidable with sufficient rig pump capacity and good mud properties (Plastic Viscosity (PV) — to carry the cutting when the mud is moving and Yield Point (YP) — when circulation is stopped temporarily with cuttings still in the annulus). In vertical wells, there are typically fewer problems. It tends to be problematic at inclinations of between 50° and 65° as cutting beds build-up on the low side of the well at anything between 40° and 75°. Counter-intuitively, it is less of a problem in horizontal wells, where the turning drill string tends to churn the cuttings on the low side of the well back into the flow-path.

Software can be used to model all these types of problems and to determine required fluid properties, pump rates for a given hole profile, size and BHA.

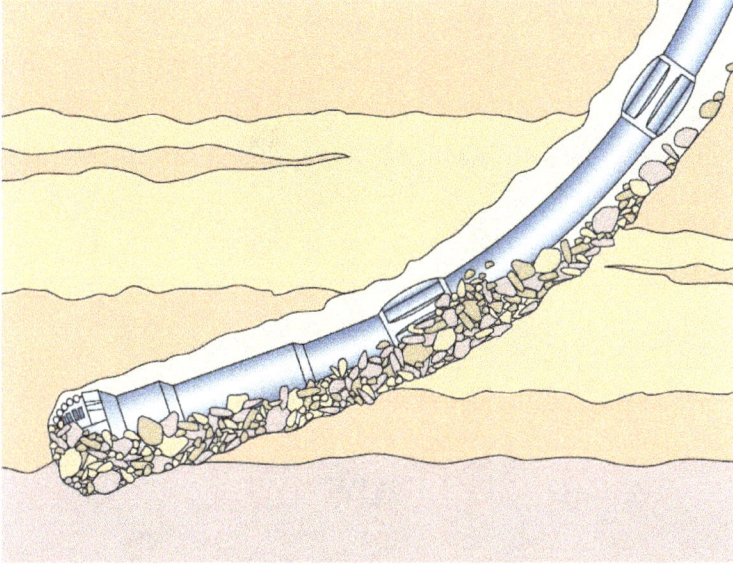

Figure 1.117. Poor hole cleaning. Illustration copyright Oilfield Review, used with permission of Schlumberger.

Chemically active formations (see Figure 1.118) are frequently drilled. These tend to be clays that react with water filtrate in drilling fluids. This causes the formation to swell into the wellbore and pinch the drill string. Where such issues are expected, inhibited water-based drilling fluids, or oil-based drilling fluids are deployed to minimise this effect. It also tends to be time-dependent — hence the focus during the drilling operation to drill, case and cement such hole sections as quickly as possible. Active formations are also managed by running the BHA through them several times to "ream" the section by scraping off any swollen clays in the wellbore.

Drill string vibration (see Figure 1.119) occurs if the drill string is allowed to impact the borehole wall. Fragile but stable formations will fracture in these circumstances and may fall into the wellbore and jam the drill string in the hole.

An **irregular borehole geometry** (see Figure 1.120) can be caused by poor BHA selection, mud system or drilling practices. The most common issues happen at the interface between formations —

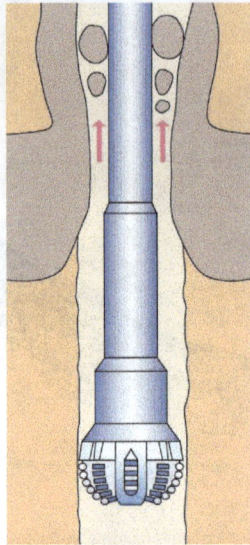

Figure 1.118. Chemically active formation. Illustration copyright Oilfield Review, used with permission of Schlumberger.

leading to ledges and other sharp features that can jam the BHA, especially on retrieval.

Fractured and faulted formations (see Figure 1.121) can fail and fall into the hole behind the bit or a stabiliser. This type of effect is aggravated by running the drill string into the hole too quickly, losses, drill string vibrations and inadequate drilling fluid properties.

Unconsolidated zones (see Figure 1.122) are typically sands found in the top hole sections of the well. They are addressed with a high circulation rate (to carry the loose sand from the well in a generally large annulus) and by using optimum mud-weight. Unfortunately, this can result in large hole diameters called **washouts**.

Key seating (see Figures 1.123 and 1.124) takes place when the drill string cuts a secondary bore in the side of the wellbore, resulting in a second lobe in the section of the well. It is caused by high doglegs (well turning too aggressively) and avoided by running a stabiliser on the top of the drill collars to open-up the hole. If detected, key

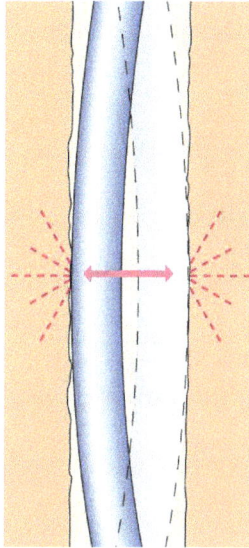

Figure 1.119. Drill string vibration. Illustration copyright Oilfield Review, used with permission of Schlumberger.

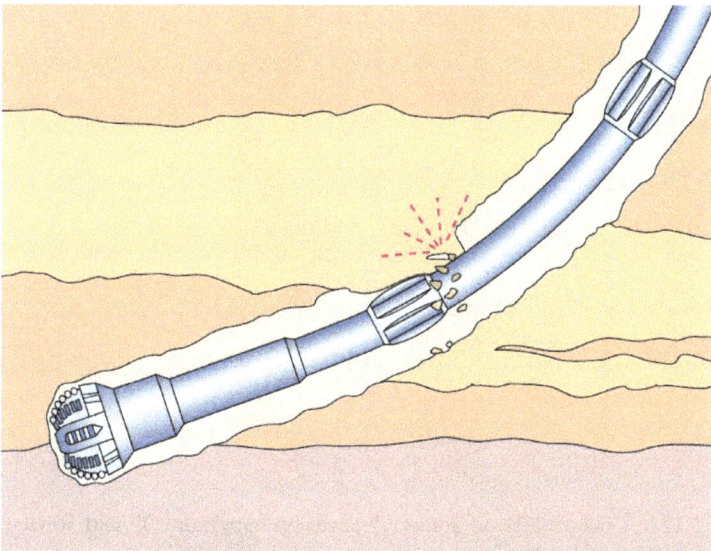

Figure 1.120. Borehole geometry. Illustration copyright Oilfield Review, used with permission of Schlumberger.

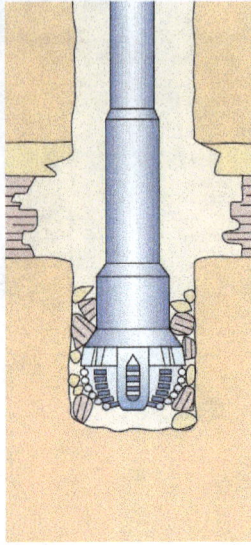

Figure 1.121. Fractured/faulted formations. Illustration copyright Oilfield Review, used with permission of Schlumberger.

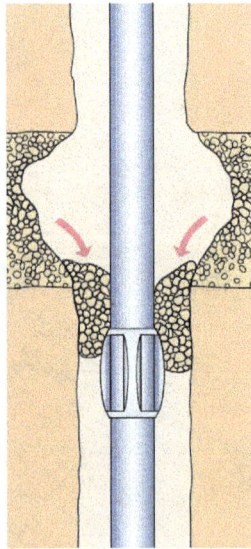

Figure 1.122. Unconsolidated zones. Illustration copyright Oilfield Review, used with permission of Schlumberger.

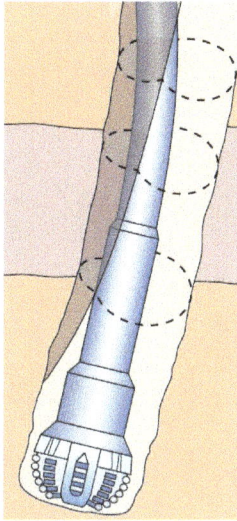

Figure 1.123. Key seating. Illustration copyright Oilfield Review, used with permission of Schlumberger.

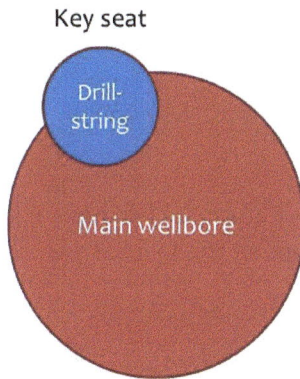

Figure 1.124. Key seating well cross-section.

seating can usually be cured by reaming the hole to smooth out the effect.

Geo-pressured formations (where the pore pressures in the formation are greater than the mud weight (see Figure 1.125)) are

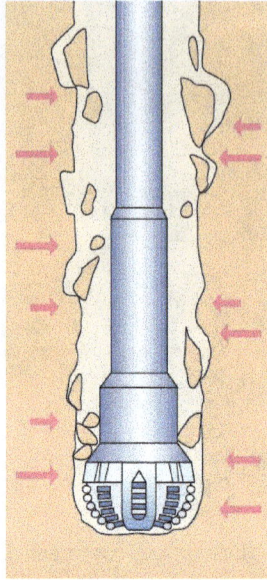

Figure 1.125. Geo-pressured formation. Illustration copyright Oilfield Review, used with permission of Schlumberger.

typically shales or claystones that have a tendency to splinter and fall into the wellbore.

Salt formations (see Figure 1.126) have a particular tendency to squeeze into the wellbore as the local stress regime is affected by the wellbore just drilled.

Cement-related (see Figure 1.127) problems can occur, particularly in the "rathole" at the base of a casing string, where cement in the open hole can easily fracture and fail.

Junk in the hole (see Figure 1.128) or **under-gauge hole** (see Figure 1.129) are generally caused by poor drilling practices such as running or pulling the BHA too quickly, dropping junk into the hole or running the cones off roller-cone bits.

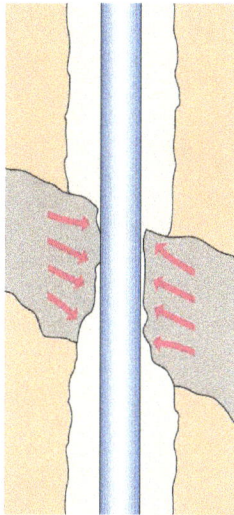

Figure 1.126. Mobile salt formation. Illustration copyright Oilfield Review, used with permission of Schlumberger.

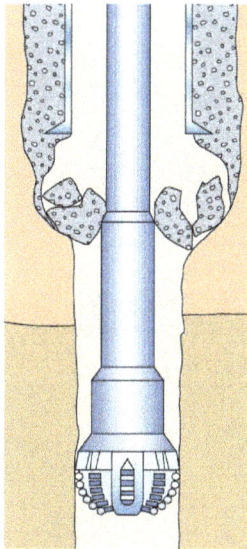

Figure 1.127. Cement related. Illustration copyright Oilfield Review, used with permission of Schlumberger.

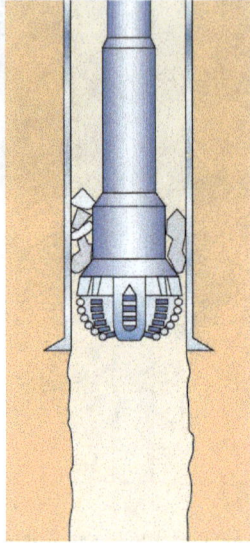

Figure 1.128. Junk in the hole. Illustration copyright Oilfield Review, used with permission of Schlumberger.

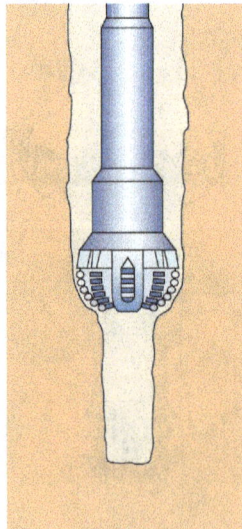

Figure 1.129. Under-gauge hole. Illustration copyright Oilfield Review, used with permission of Schlumberger.

Getting stuck can be expensive, and a single incident has been known to cost more than $100 million. Typical rig-site indications are:

- cavings (*splintery*) collected at the shale shakers — these are differentiated from *cuttings* by their larger size and curved surfaces;
- increased torque and drag;
- increased gas levels;
- circulation restricted or impossible;
- hole drilling fluid volume fill is different to that expected;
- an increase in ROP;
- cuttings and cavings not hydrated or mushy.

The first actions when sticking is detected are to:

1. Ensure circulation is maintained.
2. If the string became stuck while moving up, apply torque and jar down.
3. If the string became stuck while moving down, do not apply torque and jar up.
4. Jarring operations should start with light loading (50,000 lbs) and then systematically increased to maximum load.
5. If jarring is unsuccessful, consider acid pills, if conditions permit. Acid pills involve pumping of acid alongside the BHA and have been known to release stuck pipe.

It is better to plan to avoid "stuck pipe" and consideration at the planning stage is given to:

- design simplicity:
  - keep BHA as short as practically possible;
  - eliminate tools which are not used or have a low probability of being used.
- size drill collars/HWDP as a compromise between:
  - WOB;
  - rigidity;
  - annular clearance;
  - annular velocity across the BHA;
  - wall contact area.

- jar optimisation:
  - type of jar, placement of jar, use of 1 or 2 jars;
  - awareness of jar limitations and impact.
- dimensions:
  - accurately measure & gauge (Length, OD and ID);
    - bit;
    - stabilisers;
    - all tools in BHA;
  - ensure access to Free Point Indicator/back-off tools.
- downhole visualisation:
  - record all hole problems/issues;
  - sketch downhole situation;
  - note BHA change on tripping/drilling.
- records:
  - certification/inspection/records/operating hours;
  - remove or replace tools (jars, motors) that have reached maximum operating hours.

1.22. Fishing

Fishing is the operation of recovering an object that is:

- stuck during drilling operations;
- lost due to mechanical failure:
  - twist-off — excess torque applied to the drillstring;
  - drill-bit cones detached from the bit and left in the well;
- completion equipment being retrieved for well repair or side-track;
- junk — items accidentally dropped in the hole.

With any drill string in the hole, the driller needs to know how much force he can pull, how much torque he can apply and how much pressure he can use before the weakest part of the drill string will fail. Failure mechanisms include one or more of the following:

- twist off (pipe or connection failure due to torque);
- over-pull (pipe or connection failure in tension);

- washout (hole in the pipe eroded by pumping drilling fluid through it) — leading to failure;
- cyclic loading (metal fatigue);
- crack propagation (sometimes caused by **Sulphur Stress Cracking (SSC)** — in H_2S environments);
- mismatched components (causing local stress concentrations);
- bit cone failure;
- items accidentally dropped into the well at surface.

In summary, there are lots of pieces to get stuck, twist off, wash out, break, etc. that get left in the hole. Although here we focus on drilling, the same principles apply for tubing, casing, wireline and slick-line operations.

The **Free Point Theory** holds that the point at which the drill string is stuck, there will be no rotational nor axial strain. A tool can be run into the drill string that latches onto the inner bore of the pipe and detects very small strains when pull and/or torque is applied at surface. The tool is run until the deepest free point is detected. The next step is to apply left-hand torque to the drill string (typically 60–70% of the make-up torque but always less than that required to unscrew the weakest connection) and run a small explosive change into the string opposite the first free connection above the stuck point. The explosive is fired, the connection is "popped" open and string above the back-off point recovered to surface. For drill collars a "colliding" tool is used that creates a very powerful explosion. Casing and tubing is more delicate, and a chemical cutter is frequently used. The chemical cutter dissolves the pipe leaving a very clean cut and no debris.

Once the upper section of the BHA has been recovered, a calculation is done at the wellsite to determine the way forward — either to plug back, abandon the lower part of the well and side-track around the fish or whether to fish the lower part of the BHA. Here are some considerations at this point:

- value of the fish — RSS and MWD/LWD/geo-steering tools are expensive;
- cost of a side-track (also expensive);

- environmental considerations (e.g. if a radioactive source from a petrophysics logging tool is lost in the hole);
- condition of the upper part of the equipment to be fished;
- depth and hole condition;
- rig rate versus equipment cost;
- ...and not least, the probability of success of the fishing operations.

Retrieving the rest of the fish requires a range of specialist tools. The most expensive operations warrant the use of a specialist contractor with the experience of these operations.

Some typical fishing tools include:

The **Releasing and circulating overshot** is the most common tool used and is designed to fit over the top of the fish. It includes a spiral grapple that latches onto the OD of the fish and a seal that allows circulation down the fish. This allows full tension and torque to be applied to try to free the component.

Releasing spears are used to engage the internal bore of the fish. A feature of this tool and the overshot is that they can be released from the fish at will (usually by applying left-hand rotation).

The **Lead impression block** is a lead block run onto the top of the fish to create a physical impression in the (soft) lead. This helps the fishing engineer determine the exact profile of what they are trying to fish. For fishing in completions — where a clear brine fluid is often used in the well — a **TV camera** can be run for the same purpose.

The **Rotary taper tap** is a tool of last resort. It has a tapered screw profile that is screwed into the inner bore of the fish. As there is no releasing mechanism, a safety joint is always run above the tool to release from it if all else fails.

Grabber tools for fishing smaller items like bit cones or tools accidentally dropped into the well.

Junk baskets are used to collect very small items at the bottom by jetting the bottom of the hole by circulating quickly. The junk is lifted into the annulus where an increase in annulus cross-section causes the fluid velocity to reduce, and the junk falls into a basket that allows it to be recovered to surface.

Other tools such as **milling tools** to dress the top of the fish are available. Modern cutters can make surprisingly quick progress in milling BHA components and casing.

The fishing BHA typically comprises:

- overshot (releasing);
- bumper sub;
- drill collars;
- hydraulic jar;
- drill collars;
- accelerator;
- HWDP.

Fishing operations can take several days, and frequently the fish is recovered in parts. Care is taken to avoid making the situation worse by getting additional equipment stuck in the well. Information on internal and external diameters and profiles, lengths of components and their strength are vital for fishing and this information is always gathered before each BHA is run. A detailed record of what has been recovered is also kept, so that the hole is known to be clear once the final components are pulled out. The process can involve using the maximum forces the rig is capable of providing. BHA components recovered will always be inspected before reuse.

1.23. Formation Evaluation

In most drilling operations, there is a need to evaluate the formation being drilled. In the case of an exploration well, this will be the first opportunity to confirm the lithology and presence of fluids including hydrocarbons. This data is essential for making big decisions on whether to invest further in appraisal wells, and ultimately will inform the investment decisions required for field development.

For development wells, although the main purpose will be production from (or injection into) the reservoir, there will generally be a need to acquire data to update the geological and reservoir models (pressures and depth of contacts, etc.) and to determine the size of production tubing.

Formation evaluation information comes from the following sources:

- cuttings and flow-line data;
- continuous coring;
- petrophysical logs;
- seismic "check" shots (sonic wave velocity survey);
- sidewall samples;
- repeat formation testing tools;
- well test.

Each of these will be dealt with in this section. Details of petrophysics tools can be found elsewhere in this handbook.

1.23.1.   *Cuttings and Flow-Line Data*

It is normal to retrieve drilled cuttings at predetermined drilled intervals such as every 10 ft above the reservoir and every 1 ft in the reservoir. The cuttings are taken from the shale shaker screens and any other solids removal equipment to ensure that sampling is representative. Account needs to be taken for the lag time as cuttings are transported up the well and for any slippage estimated between cuttings and drilling fluids. The cuttings are normally washed, described at the wellsite, logged and bagged for future analysis.

Additionally, flow-line drilling fluid returns are monitored continuously for temperature, gas and other hydrocarbon shows and fluid salinity. The gas returns may be fed into a chromatograph to determine the levels of C1 (methane), C2 (e.g. ethane), C3, C4 and C5+ in the gas stream, plus any CO_2 and/or H_2S. All this information, together with basic drilling parameters and drilling fluid properties, is documented on a **mud log** that is made available to office-based staff on a daily basis.

Elements of this evaluation may be eliminated on basic development wells; however, on exploration and appraisal wells it is normal to deploy a "mud-logging" crew specifically dedicated to these activities plus the continuous monitoring of pit levels for well control safety.

1.23.2. *Continuous Coring*

Recovered core allows the direct measurement of petrophysical properties as well as detailed description of lithology and other subsurface features that may only be identifiable over a scale of inches.

Actual core drilling takes place using a core barrel, see Figure 1.130. The core barrel comprises a core bit (described in the section on bits, see Figure 1.131) which is screwed onto an outer core barrel that includes a near-bit stabiliser which is itself attached to the lower part of the BHA. There is an inner barrel that does not rotate when coring takes place, which slides over the core as it is cut. Once the coring assembly has been run into the well, a ball is dropped which diverts the fluid flow around the inner barrel to the bit. Cutting the core then takes place as smoothly as possible and avoiding pulling off the bottom. When the length of the inner barrel

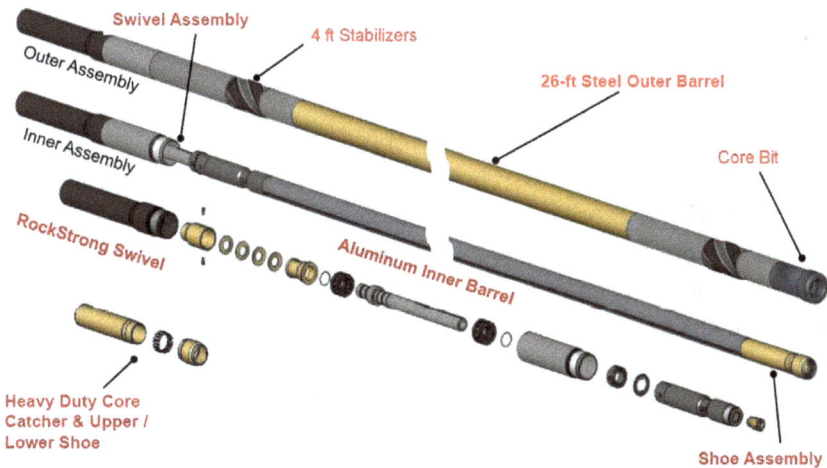

Figure 1.130. Coring assembly.

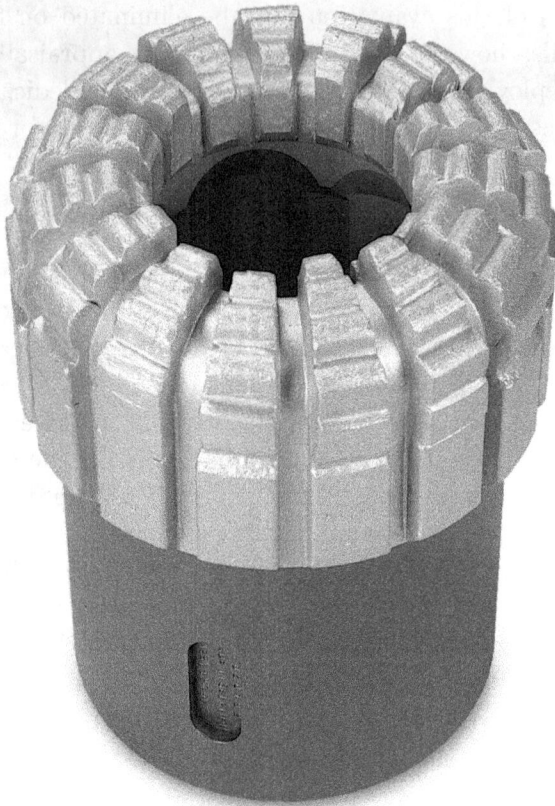

Figure 1.131. Core Head.

has been filled with the core, the bit is lifted off the bottom; the core is broken off from the bottom of the hole, and retained in the assembly with a "core catcher" arrangement. The coring assembly is then recovered to surface, the inner barrel removed and the core itself sawn up into 3 ft lengths and put into trays for further description and analysis (see Figure 1.132). Care needs to be taken in case gas is trapped in the core barrel — potentially at reservoir pressure.

The operational decision to take core (called "selecting the coring point") can be difficult. Start coring too shallow and time will be spent collecting overburden rock; start too late and the target formation will have been drilled conventionally and not cored, losing

Figure 1.132. Core in inspection trays.

information (or requiring an even more expensive side-track in which to try again). Core point detection is made based on MWD data, cuttings and drilling parameters such as ROP.

Various lengths of core can be taken at one time but there is a trade-off between length taken and probability of failing to recover it. Usually 30 ft, 60 ft, 90 ft or 120 ft cores are taken. Coring always requires additional trips to run the coring assembly, and coring ROPs tend to be very slow in order to maximise recovery of core. Core recovery (expressed as a % of rock drilled) is often 95–100% but can be much less particularly where the formation is unconsolidated.

There are several variations on coring operations. Oriented coring can be used to maintain a reference to how the core was oriented in the wellbore — this uses a combination of core barrel and directional instrumentation. Pressurised coring is another variation in which core can be maintained at reservoir pressure so that important data on the fluid content is not compromised. More sophisticated ways of protecting the core at surface — including freezing it — may be used.

1.23.3.   *Petrophysical Logs*

Wireline logging operations were first carried in the 1920s by the Schlumberger brothers, a firm that still regards wireline logging as its signature service. Wireline operations involve running a logging sonde into the well on wireline containing several electrical conductors. Typically, there are seven conductors, but it can be a single wire for some operations. The wireline conducts power to the tools plus two-way data communications, and physically supports the tools in the well.

Typically, tools are lowered into the well and logging takes place by pulling the tools out slowly while measuring one or more characteristics of the formation, casing or cement. Here is a high-level list of the most common parameters measured:

Open hole		Cased hole
• Gamma ray	• Formation dip	• Cement quality
• Density	• Seismic (sonic time to surface)	• Production pressure, temperature, flow rate
• Sonic transit time		• Casing collar location
• Resistivity (measured at various lateral depths or investigation)	• Formation pressure, temperature	• Location of fluid contacts
	• Formation fluid sample	• Perforation
• Porosity	• Formation side-wall core	• Free-point indicator
• Borehole geometry		• Severance of pipe
• Elemental analysis		
• Surface tension		

Most of these tools can be run together to optimise the use of rig time. Where multiple runs are required, it is normal to gather the most important data first, in case the well condition deteriorates. Some of the tools (density, porosity) contain radioactive sources and must be carefully managed at the surface, and not lost downhole if at all possible.

Several tools (for example, density, porosity, borehole geometry and the sampling tools) include arms that are extended out from the tool to contact the wellbore wall.

Figure 1.133. Wireline logging truck.

Wireline is run at all phases of well operations — drilling, data acquisition, casing and tubing operations, testing and production operations and to confirm well integrity. If wireline is run into a well under pressure, a wireline BOP and lubricator is rigged up that allows the tools to enter the wellbore and wireline to be run.

At the wellsite, wireline operations are managed from a wireline truck (onshore, see Figure 1.133) or a container-sized modular unit (offshore). Both are essentially the same, providing a winch with power and control system, plus electronics and computer systems to process the data, store it and transit it to the other office locations. Sheaves are rigged-up on the rig-floor and on the hook so that the tools can be easily managed. A device on the lower sheave anchor measures the tension in the cable. For subsea operation, a special sheaving arrangement compensates for rig motion.

If logging tools get stuck in the well, they can be fished like any other junk, but they are expensive so attempts are usually made to recover them, especially if they remain connected to the wireline. In this case, the fishing assembly is "stripped over" the wire. Considerable efforts are taken to recover any tool that contains nuclear sources. If left in-hole, the hole must not be drilled and cement is placed above the tool. Authorities in most countries require notification or their explicit approval for such operations.

Wireline should not be confused with slick-line which is used to run plugs and other mechanical devices in the well. Slick-line is of solid (rather than braided) construction and does not include electrical conductors.

In highly inclined wells, tools need to be assisted along the wellbore due to high levels of friction. This can be achieved either by:

1. Using a **tractor** to pull the tools along the well. This is a device that pulls itself along the wellbore by alternately setting and unsetting claws that grip the wellbore.
2. Running the wireline tools on drill pipe, in which case the cable runs inside the pipe. This is called the **Tough Logging Conditions (TLC)** technique!

Both techniques are challenging and not completely reliable.

A detailed description of the various logging tools is beyond the scope of this Chapter. Two common tools of interest are worth covering however:

The **Formation Tester** (for example **RFT** or **MDT**) is a tool used to obtain formation fluid pressure data, flow information and samples, see Figure 1.134. The tool is run to the location of interest in the well and an arm extended to press a rubber pad up against the wellbore. Then, a metal probe is extended through the filter cake and opened — up to the chamber under vacuum inside. Formation fluid is then able to flow into the chamber. Pressure, temperature and flow-rate data are obtained. The sample can also be sealed and recovered to surface. Multiple data points in the well can be taken in this way. The later (e.g. MDT) tools allow multiple samples to be kept, sample chambers to be "pumped-out" into the wellbore and include an optic sensor to determine downhole whether oil or gas has been sampled.

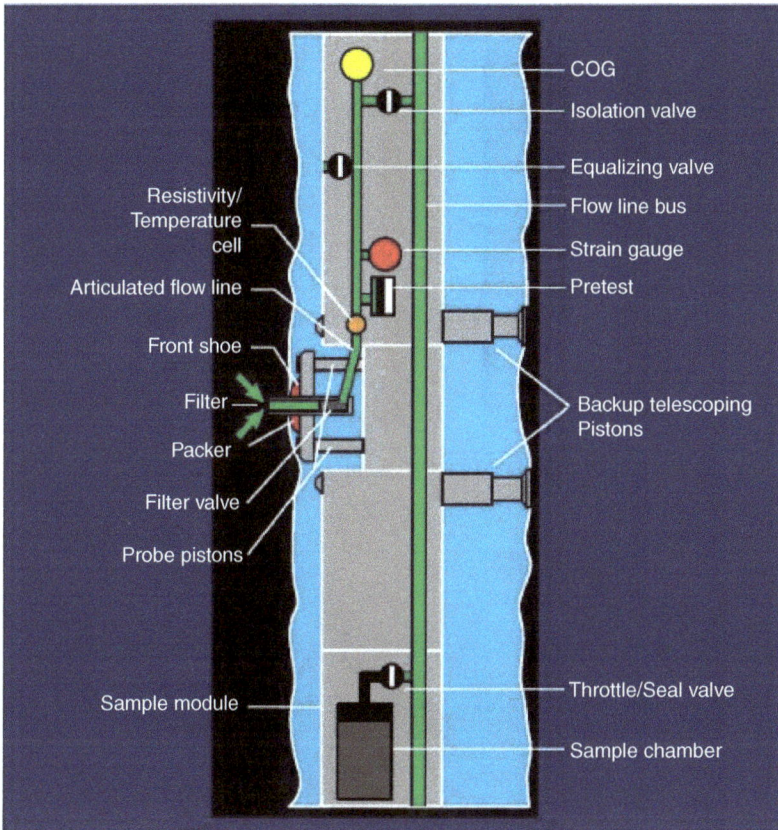

Figure 1.134. Formation Testing tool.

This tool is the only way of measuring formation pressure and permeability downhole without carrying out a well test.

The second tool of interest is the **Sidewall Coring** tool. See Figures 1.135 and 1.136. This tool comprises a set of hollow "bullets" positioned laterally in the tool, behind each of which there is an explosive charge. Each bullet is tied to the tool body with a pair of coiled retaining wires. Once the required depth is achieved, the bullet is fired and embeds itself in the formation. The tool is then raised up a few inches to pull the bullet out using the retaining wires. The process is repeated for each depth at which a sidewall core is

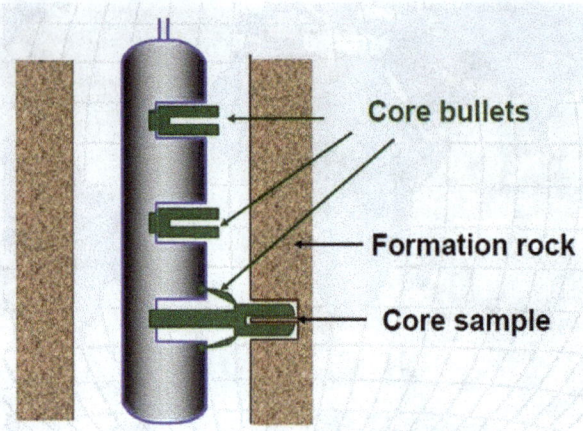

Figure 1.135. Sidewall sample tool.

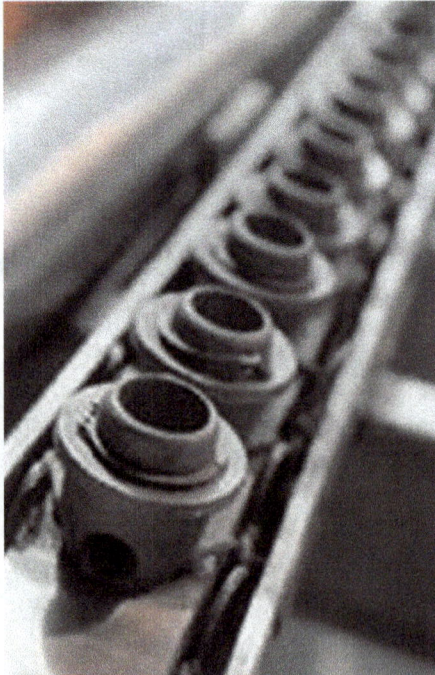

Figure 1.136. Sidewall sample tool.

required. Once finished (15 or 30 shots is normal), the tool is recovered and the sidewall cores extracted from the bullets that can be reused.

Sidewall cores are thus quick and low-cost to obtain but have the disadvantage that they are affected by near-wellbore effects of fluid invasion and are too small for carrying out permeability tests.

1.23.4. *Perforation*

In wells where the production casing and/or liner has been cemented, tubulars across the reservoir will need to be perforated. Perforating involved making holes through the pipe and cement and some distance (typically 10 cm–30 cm) into the formation, using explosive charges. The principle is to access the hydrocarbons by creating a path that does not impede flow (see Figures 1.137 and 1.138).

Perforations can be made using perforating guns on wireline, or run on tubing (called Tubing-Conveyed Perforating, TCP). The former operation is similar to any logging operation in setup; the guns are located in a depth relative to the gamma ray signature of the formation and fired with an electrical signal from surface. Very long lengths of perforating gun can be linked together and discharged simultaneously. Perforating guns are specified in terms of shots per foot, and the phasing of the shots can be varied so that for example, each is radially offset by 120°.

With TCP, the guns are run on the bottom end of the completion tubing and positioned alongside the reservoir to be perforated. They can be detonated by pressure applied from surface, or by dropping a bar inside the tubing. Once detonated, the TCP guns may be disconnected from the completion and dropped to the bottom of the well, out of the flow-path. For this reason, if TCP is intended, the well is usually drilled a few metres deeper to accommodate the spent guns.

Although possible with wireline guns, TCP has better potential for perforating underbalance — i.e. deliberately allowing the well to flow back immediately after perforating. This is thought to result in better clean-up of the perforations, and hence reduced impairment.

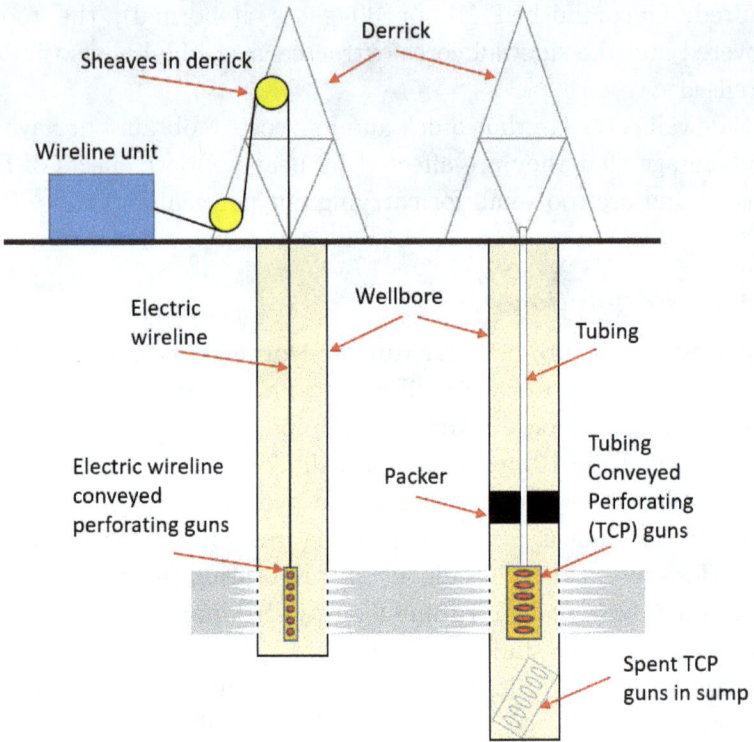

Figure 1.137. Perforating operations.

Transport, storage and use of explosives at the wellsite is carefully controlled and regulated by government authorities. Precautions are taken at the wellsite including **radio silence** (zero radio transmissions within for example 500 m of the rig) to avoid premature detonation, although newer technologies can be deployed that avoid this particular requirement.

1.24.   Well Testing

Well testing provides the following data and information:

- productivity;
- fluid properties;
- fluid composition;

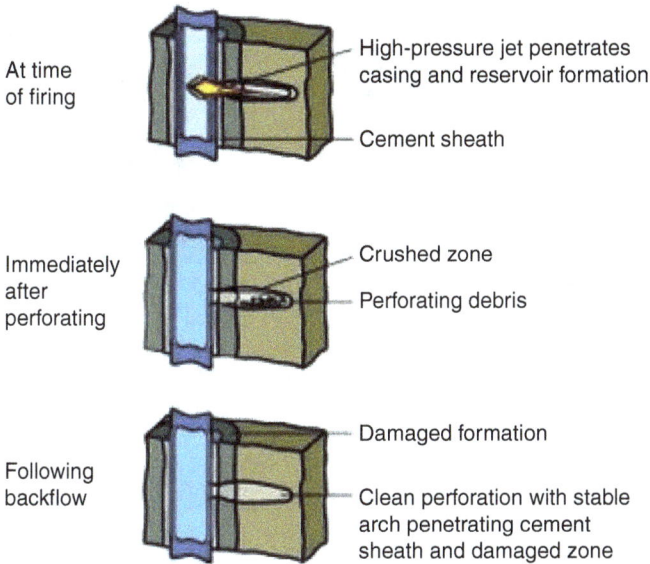

Figure 1.138. Perforating guns.

- sand potential;
- flow potential;
- pressure;
- temperature;
- combining data to prove reservoir potential, confirm well performance, and improve field productivity.

Definitive information on how well a reservoir will produce can only be gathered by conducting an actual production test (see Figure 1.139).

Production testing is usually carried out in appraisal and early development wells. It involves running a temporary completion into the well, complete with packer, Christmas tree and various tools in the test string. This may include downhole memory gauges and SSDs to allow various different zones to be perforated and produced. It may also allow for stimulation by acid or sand fraccing, as will be required in the "real" production well.

Figure 1.139. Well test flare offshore.

At surface, a temporary production system is rigged up on the wellsite to separate the gas, oil, water and solids that may be produced. A system for collecting samples is usual, and for accurately measuring the production rates of the various fluids. The hydrocarbons may be put into suitable existing hydrocarbon infrastructure or burned off (see above).

Production testing from a drillship or semi-submersible involves additional complications. The Christmas tree (on the drill floor) is fixed to the seabed, and flexible connections are required to accommodate vessel heave. Furthermore, in the event of an emergency disconnect being required between rig and seabed BOP, the test string needs to be unlatched as well. For this purpose, a **SubSea Test Tree (SSTT)** is used in the test string that includes this functionality, seals against the pipe rams in the BOP and has the ability to close-in the well using a SSSV-type valve.

Information gathered by the production test is crucial to finalising how many future production wells will be needed and the basis of their design. For example, if it transpires that the formation produces sand, this can be managed by planning gravel packed completions in the production wells. Without the production data, sand control may be omitted and the resulting workovers to install it would be very costly indeed and would certainly delay production, project payback and damage the overall economics of the project.

1.25. Fraccing and Stimulation

Fraccing and stimulation is a whole subject on its own. Here, we will focus on the basic practical issues.

There are a variety of different fracture and stimulation options. **Hydraulic fracturing** (called **fraccing**) is often applied in a well — usually horizontal — that intersects natural fractures in the reservoir formation (see Figure 1.140). Ultimate recovery and the rate of gas or oil production from these wells can often be increased by opening-up the fractures and propping them open with sand grains. This allows a greater volume of reservoir to be accessed by extended and additional fractures, and lower pressure drops along the fractures leading to greater production rates. Fraccing is accomplished by pumping water at high pressure into fractures, followed by sand suspended in viscous water (though in reality the actual sequence and recipe of pumped fluids is more complex). This treatment is most effective when applied in horizontal sections of the well, and techniques exist to efficiently treat and move from one section to another. When put on production these wells flow back most of the injected fluid (but the sand remains in the fractures).

The surface rig-up for fraccing operations on land comprises many sand blending and pumping units lined-up to a special "fraccing tree" and controlled from a central cabin at the site. Figure 1.141 illustrates the scale of a typical site setup. Very high injection pressures (up to 15,000 psi) can be used, and huge quantities (hundreds of tonnes) of sand can be pumped away into any one well.

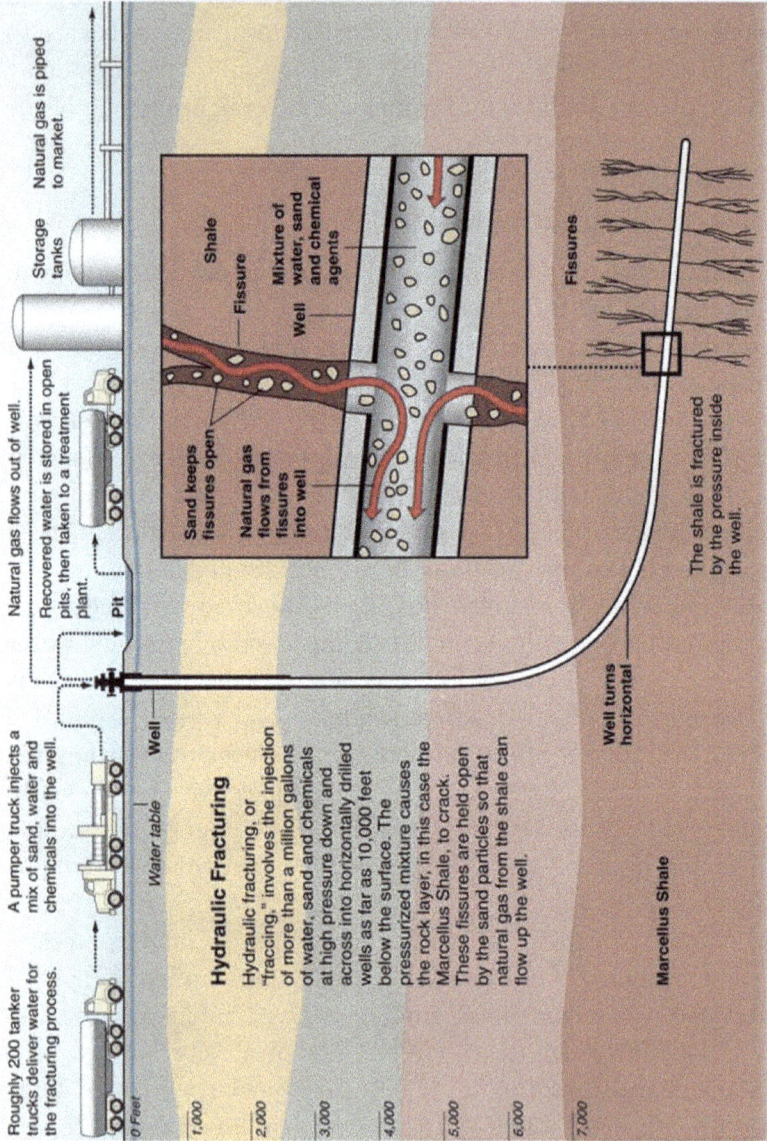

Figure 1.140. Fraccing operations.

Figure 1.141. Fraccing operations at the wellsite.

The fraccing operation can cost as much as the drilling of shale gas wells.

Acid fraccing tends to be used in limestone rather than sandstone reservoirs, and far smaller volumes are pumped. Hydrochloric or nitric acid is applied to fractures in the limestone, which partially dissolves the formation opening-up the hydrocarbon flow-paths.

Fraccing has become a controversial operation, particularly in shale gas activities in the USA. This centres on the risk of groundwater contamination due to the fractures penetrating shallow drinking-water aquifers and/or poor cementation and/or poor maintenance of well integrity of its lifetime. In reality, there have been very few genuine examples of these problems, which are entirely avoidable. In the UK, there have been examples of very small-scale earthquakes (2–3 on the Richter scale). France and Bulgaria have banned fraccing.

Public perceptions are important and the industry needs to do a better job of managing concerns. Several service companies market fraccing chemicals based on food additives to help address the issues.

1.26. Well Control and Blowouts

We have touched on various aspects of well control already in this chapter. To recap, the main reasons for well control issues are:

- **Failure to keep the hole full of fluid (of correct density) during a trip** — this is the responsibility of the Driller, and he uses the trip-tank — a sensitive and instrumented mud pit — to ensure that the hole is taking or returning exactly the correct quantity of fluid that relates to pipe steel displacement.
- **Swabbing while tripping** — can occur if the Driller pulls tubulars out of the hole too quickly. This causes a temporary reduction in bottom-hole pressure, potentially to a value lower than the pore pressure, which creates an influx of fluid (such as gas, oil or water) into the bottom of the well. Swabbing is prevented by pulling out of hole slowly, pumping out of hole using a top-drive, maintaining the mud in good condition, wiping the hole to treat any tight spots and avoiding BHAs with too many stabilisers.
- **Lost circulation** — can result in loss of hydrostatic head in the well if it is not kept topped-up with fluid of suitable density. Lost circulation is avoided by reducing circulation rate and using LCM in the drilling fluid.
- **Insufficient density of fluid** — avoided by regular testing of the drilling fluid and minimising levels of entrained gas.
- **Abnormal pore pressure** — is not totally avoidable. However, it can be managed by proper well planning using all subsurface knowledge available, and by using a suitable mud weight which offers a margin to cover the unexpected. There are techniques such as calculating the D-exponent that purport to predict pore pressures as the well is drilled. Where pore pressures are uncertain, it is especially important that the drill crew be vigilant for indications of kicks such as flow, increases in ROP, changes in torque etc.
- **Drilling into an adjacent well** — happens occasionally but is totally avoidable if correct surveying, drilling and well isolation procedures are followed. A more insidious risk arises when shallow formations have been charged, possibly by leaks from other wells.

If all these issues are not managed correctly, the effect can be devastating···

A blowout may be defined as uncontrolled flow from a well — either to the external environment or internally from one formation to another. The former are the most dramatic and have the potential to kill or injure anyone close to the wellsite and cause significant environmental damage.

In the early years of the oil industry, for example at Spindletop in 1902 (see Figure 1.142), a "successful" well was one where the oil blew out uncontrolled from the well! This was then collected and processed.

The highest profile blowout in recent years was on the BP-operated Transocean Horizon rig that was suspending a well in the Macondo field in the US Gulf of Mexico in April 2010 (see Figure 1.143). Eleven offshore crew were killed, and there was very significant environmental damage from the resulting oil spill from

Figure 1.142. Spindletop.

Figure 1.143. Transocean Horizon/BP Macondo Blowout.

uncontrolled flow from the well over several weeks. BP's reputation and market value was devastated, and there were lawsuits and prosecutions involving companies, US state and federal agencies. The loss of assets and setback in developing the field have been considerable, and the clean-up cost billions of dollars.

For a blowout to happen two things must occur:

What fails?	Why it fails?
The **primary barriers** fail This is usually loss of hydrostatic pressure provided by drilling mud, wet cement or completion fluid that (over) balances the reservoir pressure	• Human error (on the rig, but also in the design phase) ○ Lack of awareness ○ Failure to follow procedures ○ Lack of skills ○ Failure to communicate ○ Swabbing during drilling and workover operations

(Continued)

(Continued)

What fails?	Why it fails?

The **secondary barriers** fail — drill pipe safety valve, BOP, wellhead, valve, etc	• Equipment failure
	  ○ Design
	  ○ Maintenance
	
	• Human error
	  ○ Failure to follow procedures
	  ○ Lack of knowledge
	  ○ Lack of skills
	  ○ Inability to manage the crisis

Blowouts can be analysed using the Swiss Cheese model (see Figure 1.1). In the case of Macondo, the following may have contributed to the incident:

- inadequate cement bond around the casing due to poor slurry design, testing and inadequate use of liner centralisers;
- failure of float shoe and collar;
- well incorrectly inflow tested;
- failure of BOP system to close/shear pipe when well flowing;
- failure of riser to disconnect;
- failure to divert flow at surface;
- failure that allowed gas ingestion into diesel power generation units on rig;
- failure to stop the operation when concerns were raised;
- failure of the regulatory system to adequately manage offshore safety.

There were many and varied contributions, the absence of any one of which would have prevented the full incident. Importantly, all industry players must learn from this and other incidents to avoid repetition and reduce risks further.

What needs to be emphasised is the significance of human factors in this and all other well control incidents. Many of the above reflect lapses of judgement, inadequate communication and a poor understanding of risks. These "softer" issues are the focus of several current industry initiatives that build on Crew Resource Management (CRM) practices in the airline and medical industries.

1.27. Underbalanced/Managed Pressure Drilling

Earlier in this chapter it was mentioned that at all times during drilling operations the fluid in the well must exceed the pore pressure in the open and exposed wellbore. Recent technological advances have enabled a technique called variously Managed Pressure Drilling (MPD) and Under-Balanced Drilling (UBD). These techniques (essentially the same) deliberately reduce the hydrostatic head in the well to match or underbalance the pore pressure in the open hole. This has the following potential benefits:

1. Reduced impairment of the reservoir (and hence better productivity), due to less invasion by the drilling-fluid.
2. Increased information from the reservoir (in terms of productivity data and samples) as it is drilled, which can be used to make real-time decisions.
3. Faster rates of penetration as the pressure holding down the rock as the bit cuts through it (called the chip hold-down pressure) is reduced or rendered negative.
4. Drilling in areas where the margin between formation pore pressure and fracture gradient is very narrow — such as in HPHT regions.
5. Better well control arising from a closed-circuit drilling fluid circuit, top-quality instrumentation and very skilled crew.

In most drilling operations, the mud flow at surface is at atmospheric pressure. With MPD, a choke is used on the annulus before the drilling fluid reaches the shale shakers. This enables the bottom hole pressure to be altered without adjusting the mud density. By using a lower-than-usual fluid density, the backpressure on the well can be reduced at the choke to exactly balance or underbalance the pore pressure. This situation can be maintained using a control system that detects increases in surface volume. Key requirements are:

- A choke that manages the backpressure while allowing drilled cuttings to pass.
- Flow meters in inlet and outlet flowlines that can provide indications of well flow.
- A rotating head that allows the drill string to enter and pass through the BOPs under pressure.
- A reliable (and failsafe) control system that manages the entire process.
- Skilled UBD operators who work closely with the Driller and other rig crew.

1.28. Drilling and Completion Fluids

Drilling fluid is a very important part of drilling and completion activities. Failure to maintain drilling fluid in good condition is the primary cause of non-productive time (arising from stuck pipe).

1.28.1. *Drilling Fluids*

The functions of drilling fluid (and of the completion fluid — denoted an asterisk *) are to:

1. **Remove cuttings from well:** The fluid properties — density, velocity and rheology — need to be suitable for lifting the cuttings up the annulus, meanwhile preventing them swelling or breaking into small pieces.
2. **Suspend and release cuttings:** If circulation stops, cuttings in the annulus need to be suspended and prevented from dropping back down the hole where they may jam the BHA — so high gel strengths are required. At surface, the mud must release the cuttings at the solids removal equipment, and gas in the degasser.
3. **Control formation pressures*:** This is the primary barrier in the well when drilling. The drilling fluid includes weighting material such as barite or haematite to increase its density. This

density ensures that the drilling fluid pressure at all points in the well exceeds the pore pressure at that depth.

4. **Seal permeable formations:** This is an important property because if drilling fluid is lost into the formation, more needs to be added at surface. The fluid then costs more, and any well control issues are more difficult to recognise and manage. The fluid is designed so that newly exposed borehole wall is coated by very fine solids suspended in the mud (called a mud cake). During the process, a very small amount of fluid (called mud filtrate) flows into the formation. The mud cake is quite hard and impermeable and generally prevents any further losses.

5. **Maintain wellbore stability:** Wellbore stability is achieved when a mud weight is selected that minimises disruption to the *in situ* rock stresses. This can vary with well inclination and azimuth.

6. **Inhibit reactive formations:** Claystone, salt and other formations can react to mud and/or mud filtrate by swelling to several times their original volume if unconstrained. This can result in drilling problems such as poor circulation and "stuck pipe". Water-based mud in particular needs to be inhibited — sometimes the mud is based on a salt such as potassium chloride (KCl) to provide this.

7. **Minimise formation damage*:** Given that the purpose of most wells is to produce hydrocarbons or inject water, it is essential that future productivity of the reservoir section not be damaged. The most common types of damage are:

   - mud or drill solids invade the formation matrix, reducing porosity and permeability;
   - swelling of formation clays within the reservoir matrix reducing permeability;
   - precipitation of solids due to mixing of mud filtrate and formation fluids;
   - mud filtrate and formation fluids forming an emulsion.

   Sometimes a drill-in fluid is used — with characteristics specifically designed for the reservoir. For completion fluids, filtered brines are frequently used to maintain overbalance without impairing the formation.

8. **Cool and lubricate the bit and drilling assembly:** As the bit cuts rock, it can become very hot due to the friction. Excessive heat would result in failure of roller-cone bearings. As the BHA rotates and slides in the wellbore, friction needs to be minimised, especially in high-inclination wells.

9. **Transmit hydraulic energy to mud motor/turbine:** Where a downhole mud-motor or turbine is used, the power is transmitted to the tool via the drilling fluid. The fluid column also needs to support MWD/LWD mud pulse signals without interference.

10. **Transmit hydraulic energy to the bit:** The cutting efficiency of the bit is determined by how well the bottom of the hole is cleared of cuttings (to avoid redrilling them), the cleaning of the cutters and the jetting action of the fluid. These are all essential characteristics of the drilling fluid.

11. **Ensure adequate formation evaluation:** This depends on:

    - Low fluid loss to minimise the flushing away of hydrocarbons near the wellbore which would reduce the accuracy of logging tools, and avoiding thick wall cake that could result in stuck logging tools.
    - Enabling the drilling of in-gauge hole so that logging tools work correctly.
    - Optimising overbalance to avoid the risk of differential sticking using certain tools.

12. **Minimise corrosion*:** Corrosion can lead to loss of casing integrity (especially opposite aquifers) and loss of drill string integrity, e.g. twist offs. There is also a need to control the effects of oxygen and hydrogen sulphide corrosion using hydrated lime and amine salts (to maintain pH). Oil-based mud has excellent corrosion inhibition properties.

13. **Facilitate cementing and completion:** Just before cementing the well the mud needs to be conditioned to provide suitable properties for efficient mud removal — i.e. low yield point and

gel strength. This should ensure the cement displaces the mud from the hole and is not contaminated by it.

14. **Minimise impact on environment***: There is a great deal of focus on minimising environmental impact from the use and disposal of drilling and completion fluids, cuttings impregnated with these fluids and containers used for their transportation. Here are some focus areas:

- avoiding the use of OBM and SBM (see below);
- if OBM or SBM must be used, deploying containment systems rather than for example depositing cuttings onto the seabed when offshore;
- ensuring cuttings cleaning and capture is deployed if required;
- returning cuttings for recycle processing if required;
- drilling smaller holes.

All these characteristics must be retained at high temperatures and pressures, and sometimes at low temperatures (e.g. when mud is stationary in the well at the seabed or in the pits in arctic conditions).

There are three common types of drilling fluid:

1. Water-Based Mud (WBM):

   - water-base with clays (bentonite) and other chemicals;
   - lowest cost, but some formations react to water in the filtrate.

2. Oil-Based Mud (OBM):

   - base petroleum product, e.g. diesel fuel;
   - toxic;
   - very good drilling/formation properties;
   - medium cost.

3. Synthetic-Based Fluid (SBM):

   - base synthetic oil;
   - less toxic;
   - very good drilling/formation properties;
   - expensive.

1.28.2.   *Additives*

The following **drilling fluid additives** are typically used to manage mud properties:

Additive	Purpose
Alkalinity and pH control	Control the degree of acidity or alkalinity of the drilling fluid. Most common are lime, caustic soda and bicarbonate of soda
Bactericides	Reduce the bacteria count. Paraformaldehyde, caustic soda, lime and starch preservatives are the most common
Calcium reducers	Used to prevent, reduce and overcome the contamination effects of calcium sulphates (anhydrite and gypsum). The most common are caustic soda, soda ash, bicarbonate of soda and certain polyphosphates
Corrosion inhibitors	Control the effects of oxygen and hydrogen sulphide corrosion. Hydrated lime and amine salts are often added to check this type of corrosion. Oil-based muds have excellent corrosion inhibition properties
De-foamers	Used to reduce the foaming action in salt and saturated saltwater mud systems, by reducing the surface tension

(Continued)

(Continued)

Additive	Purpose
Emulsifiers	Create a homogeneous mixture of two liquids (oil and water). The most common are modified lignosulphonates, fatty acids and amine derivatives
Filtrate reducers	Used to reduce the amount of water lost to the formation. The most common are bentonite clays, CMC (sodium carboxy-methylcellulose) and pre-gelatinised starch
Flocculants	Cause the colloidal particles in suspension to form into bunches and solids to settle out. The most common are salt, hydrated lime, gypsum and sodium tetraphosphates
Foaming agents	Most commonly used in air drilling operations. They act as surfactants, to foam in the presence of water
Lost Circulation Materials (LCM)	Inert solids are used to plug large openings in the formations, to prevent the loss of whole drilling fluid. Nut plug (nut shells), and mica flakes are commonly used
Lubricants	Used to reduce drillstring and bit torque by lowering the coefficient of friction. Certain oils and soaps are commonly used
Pipe-freeing agents	Spotting fluids in areas of "stuck pipe" to reduce friction, increase lubricity and inhibit formation hydration. Commonly used are oils, detergents, surfactants and soaps

(Continued)

(Continued)

Additive	Purpose
Shale-control inhibitors	Control the hydration, caving and disintegration of clay/shale formations. Commonly used are gypsum, sodium silicate and calcium lignosulphonates
Surfactants	Reduce the interfacial tension between contacting surfaces (oil/water, water/solids, water/air, etc.)
Weighting agents	Provide a weighted fluid higher than the fluid's specific gravity. Materials are barite, hematite, calcium carbonate and galena

1.28.3. *Fluid Properties*

It is clear from the above that managing the performance of the drilling fluid is a very important part of managing drilling operations at the wellsite. For the more expensive and demanding wells, a "mud engineer" is frequently employed dedicated to this task. He will typically ensure that mud-weight and rheological properties are measured at least every hour when drilling. Other tests are done twice a day. The following lists the main parameters measured and reported:

- Density mud weight in ppg (lbs per US gallon) or equivalent gradient in psi/ft.
- Viscosity, Plastic Viscosity (PV), Yield Point (YP) and gel strength after stationary periods of 10 seconds and 10 minutes. These measurements are made using a Fann viscometer.
- Fluid loss/filter cake thickness.
- pH.
- Solids content.
- Sand content.

- Ca^{2+} (concentration of calcium ions).
- Oil/water ratio (for OBM and SBM) and many others.

In agreement with the company representative, the mud engineer will condition/treat the mud to the specifications in the drilling programme.

1.28.4.   *Completion Fluids*

The requirements of completion fluids are simpler than those of drilling fluids. See above list where completion fluid requirements are marked thus*. In particular, the completion fluid must maintain overpressure on the reservoir formation but must not impair its future production. These requirements mean that very clean fluids are used — typically filtered brines including seawater, calcium chloride ($CaCl_2$), calcium bromide ($CaBr_2$) and zinc bromide ($ZnBr_2$). Depending on the preference of the Operator, these fluids may, or may not, be left in the well once completed.

The completion fluid in the annulus of the well during production must protect casing and tubing from corrosion, minimise pressure differential across packer and casing and remain pumpable. These properties must not degrade over time in a high temperature environment.

1.28.5.   *Cuttings and Waste Management*

In many locations, particularly offshore, dumping cuttings into the sea has become environmentally controversial. In many cases, there are government limits of the level of oil-on-cuttings that are disposed of on the seabed (such as $<5\,g$/per kg of dry cuttings) that have proven difficult to meet with existing technology. In other areas such as Norway, no oil can be discharged on cuttings.

The industry has met these requirements by:

1. Developing cuttings drying devices such as heaters that vaporise the oil off the cuttings and condense it back in a collection device for reuse. The cuttings are then deposited on the seabed,

2. Grinding cuttings into a slurry and reinjecting them down the annulus of a suitable well into a benign but permeable formation, and

3. Developing "skip-and-ship" infrastructure to collect cuttings and transport them back to shore for environmentally acceptable processing and disposal.

1.29. Well Workover and Maintenance

Most of this chapter is focused on new well construction — i.e. drilling. As mentioned in Section 1.3, well workovers and maintenance are also important areas of well engineering and can be required at any time during the life of the well.

Workovers typically have the following objectives:

- **Stimulate additional production from the reservoir** — for example, by stimulation or reperforation, probably after running logs such as the TDT (Thermal Decay Time) wireline logging tool to confirm hydrocarbon contacts.
- **Close off production (e.g. water) that may not be desirable** — this may be done by running wireline plugs, shifting SSDs or setting cement plugs.
- **Clean-out wells to remove scale or other debris** — this may require running scraping tools on slick-line, pumping an acid wash down the tubing, or ultimately recovering and re-running completion tubulars. Naturally-occurring radioactive scale, which can be deposited in production tubulars over an extended period of production, is potentially a high risk in these operations, and needs to be planned for and managed safely.
- **Various activities in difficult-to-enter wells** — some wells such as subsea wells require a complicated (and expensive) re-entry process using a drilling rig or wireline operation either from a workboat (see Figure 1.144) or, in future, a subsea wireline re-entry system.
- **Repair physical wellbore problems** — a multitude of different operational problems can occur such as sand ingress, corrosion of tubulars, reservoir compaction leading to casing and/or tubing

Figure 1.144. Helix Subsea intervention vessel the Well Enhancer.

failure and many others. These generally require significant rework, and a genuine option will exist to abandon such a well and respud a new one.

• **Prepare a well for side-tracking operations** — towards the end of well life, it can make economic sense to side-track the existing well to a new production location. Preparing the well for side-track is akin to a workover.

The combination of unique wells and the large variety of activities listed above results in a wide variety of operations. These may require rig, snubbing unit, coiled tubing unit, wireline or slick-line activities, or more likely a combination of several of these. Programmes for these activities need to recognise that surprises are likely to be uncovered during the operation and therefore will include many contingencies.

Maintenance is required to ensure that well integrity is ensured. This includes monitoring all annulus pressures and being aware of any leaks in wellhead or casings, fluids in each of the annuli as well as maintaining and lubricating valves, actuators etc. Clearly documented well status information, a maintenance programme,

KPIs to measure performance, a reporting system and clear account-abilities are essential.

There are several examples where well integrity and annulus pressure management was not given the attention it deserved, with consequences including uncontrolled flow of hydrocarbons to surface.

1.30. Well Abandonment

At the end of the life of a well, it needs to be abandoned so that future flow is not possible to surface or between individual formations. For exploration wells, this occurs right after drilling and evaluating the prospective formations, whereas for development wells it might be 50 years before they are abandoned, when economic production is no longer possible. Abandonment generally requires use of a drilling rig to recover heavy casings and wellheads.

Abandonment usually involves removing the completion, setting cement plugs across production and hydrocarbon-bearing zones, cutting and retrieving casings and wellheads and setting "T-shape" cement plugs across these casing cuts. Such plugs will generally need to be inflow tested.

Each company may have standards covering this or may refer to an industry standard such as that depicted in Figure 1.145 taken from guidelines set by the Oil and Gas UK organisation.

1.31. Artificial Lift

Completion string components have been covered in Section 1.15. Here we'll summarise the main types of artificial lift and how they impact well design.

Artificial lift is used to enable or enhance production from low-pressure reservoirs. Frequently, during the production lifetime of a field the reservoir pressure decreases; artificial lift can be installed in the original well in anticipation of this, or retrospectively as required by working over the well. Artificial lift is only applicable for oil wells (which might produce some gas and water too, of course).

The following types of artificial lift will be covered here:

- beam pumps;
- Progressive Cavity Pumps (PCPs);

Permanent Abandonment Barrier schematic
"Restoring the Cap Rock"

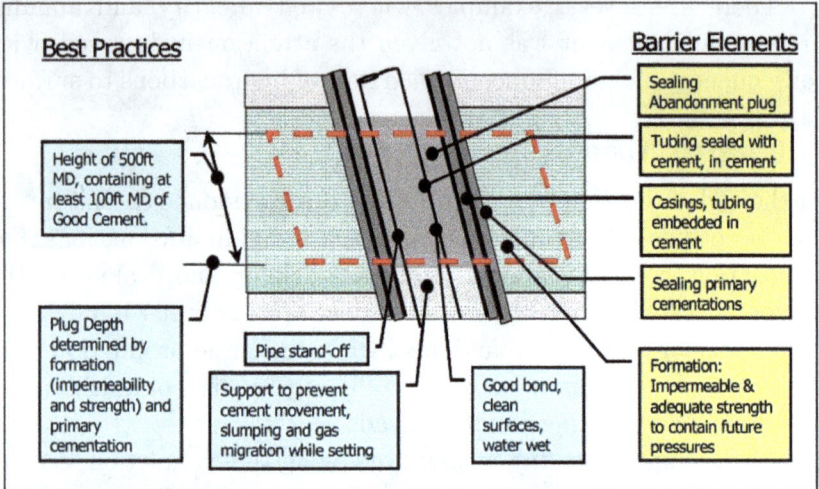

Best Practices

Barrier Elements

Sealing
Abandonment plug

Tubing sealed with
cement, in cement

Height of 500ft
MD, containing at
least 100ft MD of
Good Cement.

Casings, tubing
embedded in
cement

Sealing primary
cementations

Plug Depth
determined by
formation
(impermeability
and strength) and
primary
cementation

Pipe stand-off

Support to prevent
cement movement,
slumping and gas
migration while setting

Good bond,
clean
surfaces,
water wet

Formation:
Impermeable &
adequate strength
to contain future
pressures

Oil & Gas UK

Figure 1.145. Extract from OGUK abandonment guidelines.

- Electrical Submersible Pumps (ESPs);
- gaslift.

1.31.1.   *Beam Pumps*

Beam pumps (see Figure 1.146) are the simplest and most common
form of artificial lift, and are mostly used for shallow, low-production
onshore wells. They comprise a downhole pump, connecting rods
(called sucker rods) to surface and the surface beam pump as
illustrated in Figure 1.147. The downhole pump consists of a piston
driven up and down a cylinder. Valves in the piston and the cylinder
control the inflow and lifting of the oil up the production tubing. Any
gas separates from the oil at the bottom of the well and is produced
up the annulus.

Inside the tubing, the sucker rods reciprocate to drive the pump.
Sucker rods are solid and are screwed together in sections like very
small drill pipe — (see Figure 1.148).

Figure 1.146. Beam pumps.

At surface the beam pump connects to the sucker rods via a "stuffing box" and polished rod that prevents leakage of oil.

Beam pumps can be adjusted for stroke length and speed according to the lift characteristics of the well. They may operate intermittently. Instrumentation on the pumps can detect when the pump or sucker rods are about the fail — the well can then be worked over before failure actually occurs. The beam pump is highly tolerant of sand and water in the oil; capital costs are low but maintenance costs can be quite high.

1.31.2. *Progressive Cavity Pumps (PCPs)*

The progressive cavity pump works under the same principle as the "mud motor" described elsewhere in this chapter. It comprises a smooth lobed shaft with a twisted profile that sits into a rubber stator with one more lobe than the rotor (see Figures 1.149 and 1.150). The rotor is turned from the surface using a motor/gearbox and control the system, and torque is transmitted to the PCP via sucker rods. The oil flows up the tubing, and gas flows up the production annulus. PCPs are tolerant of water and sand.

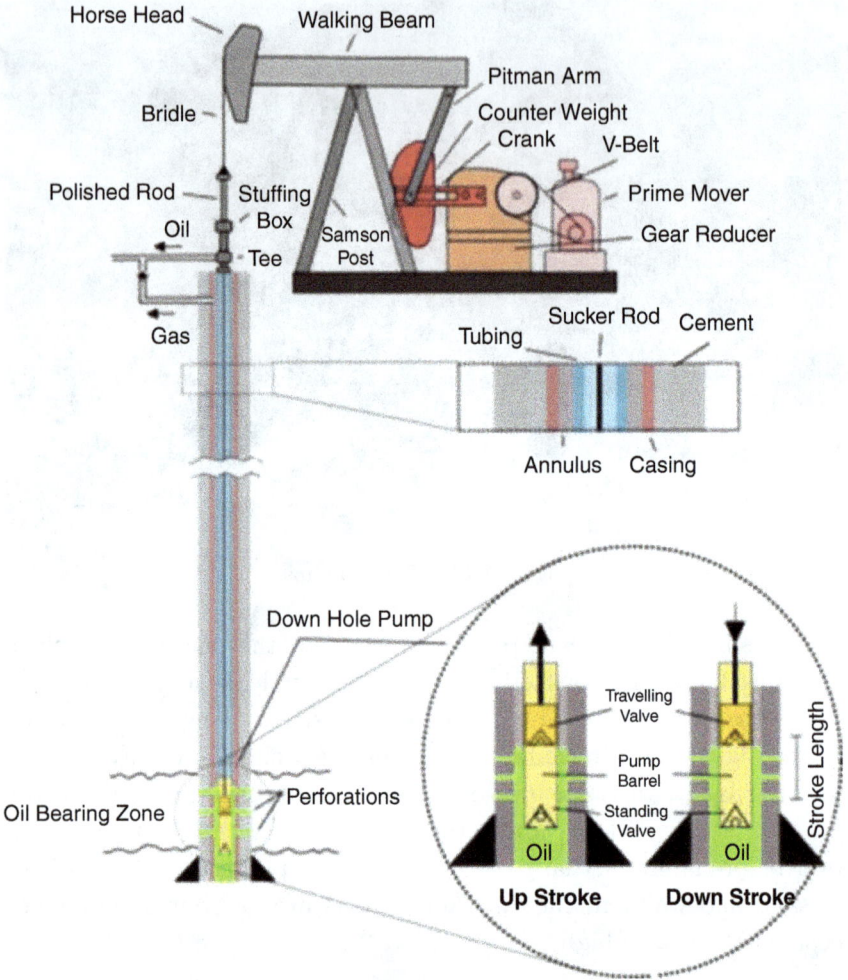

Figure 1.147. Beam pump details.

1.31.3.   *Electrical Submersible Pumps (ESPs)*

ESPs are most commonly applied to medium and high-production wells offshore and onshore. The ESP is powered by an electric motor which is encapsulated with the pump and suspended from the tubing at the bottom of the well. An electric power cable is run from a surface controller, through the wellhead and down the

Figure 1.148. Sucker rods.

outside of the tubing to the motor which sits below the pump (see Figures 1.151–1.153). ESPs are less tolerant of sand and water than other types of pumps. ESPs are sized to the production capability and characteristics of the well. ESPs may fail for many reasons including pump, motor or cable failure, sand or excessive wear. An important consideration is **Mean Time Between Failure (MTBF)**, which indicates how often the pump will need to be changed. Over recent years, MTBF has improved and today 3+ years is common.

Like bean pumps and PCPs, ESPs completions do not normally facilitate running logs across the reservoir when installed.

1.31.4. *Gas Lift*

Gas lift systems rely on the injection of gas into the production column to reduce the hydrostatic pressure and increase vertical flow. Single-string and dual-string gas lift completions are shown in Figure 1.154.

The lift gas is injected down the annulus and flows into the tubing via a side pocket mandrel (described in Section 1.15.8). As

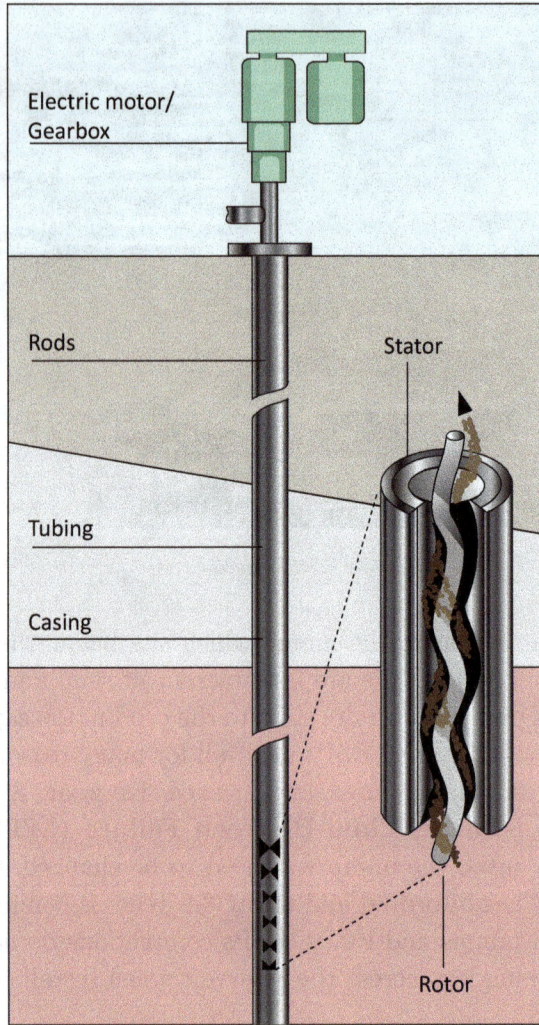

Figure 1.149. PCP general arrangement.

gas is supplied to the annulus the uppermost valve opens. The upper section of the tubing is then unloaded. Once the top part of completion is producing, the uppermost gas lift valve closes. This sequence continues until only the lowermost valve injects gas into the

Figure 1.150. PCP rotor/stator.

tubing, and this is the operating mode. The injected gas is recovered from the oil at the separation stage, after which it is compressed for re-use.

The gaslift design, setting depths of the mandrels and pressure settings on the mandrel valves are all important points required to make the gas lift completions efficient. Unlike most other artificial lift arrangements, full-bore access remains in gas-lifted wells, allowing logging operations and other activities such as reperforating, operating SSDs etc.

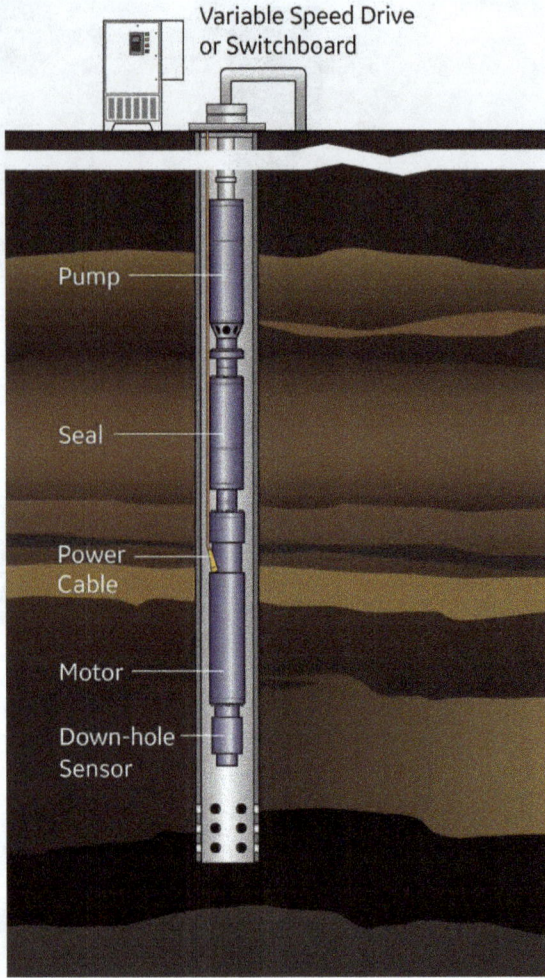

Figure 1.151. ESP downhole components.

1.32. Other Completion Designs

Multi-lateral wells. It can sometimes be more cost-effective to drill a single bore through the overburden and use multiple legs in the reservoir(s), and there are various alternative configurations (see Figure 1.155). Figure 1.156 shows a typical example. Installing completions into these wells is complicated, in particular running the

Figure 1.152. ESP power cable.

second leg through the junction, particularly if this junction must seal off both wellbores from the formation. Future maintenance of the well such as running production logs requires "selectivity" — to ensure that the tools enter the intended wellbore.

Figure 1.157 shows an extreme but realistic multi-lateral well superimposed to scale on a photograph of Rio de Janeiro.

1.33. Well Engineering Organisation and People

This chapter has focused on equipment, procedures and techniques used in well construction. However, people are a very important aspect of safe and efficient operations. At all levels, competent, motivated people in many organisations need to work together to safely drill, complete and maintain wells. Here, we will look at some of these roles:

Rig-based organisation

On a typical drilling rig, as illustrated in Figure 1.158, there will be people working for several different organisations, as follows:

1. The **Drilling Contractor** (rig owner) will employ a team of people to manage the manual operation of the rig itself, and cover maintenance.

Figure 1.153. ESP pump stages.

2. The oil/gas **Operating Company** operator is the company who has leased the rig to drill the well.

3. **Service Companies** are organisations that provide specific skilled people and/or equipment on the rigsite, and mostly work for the Operating Company.

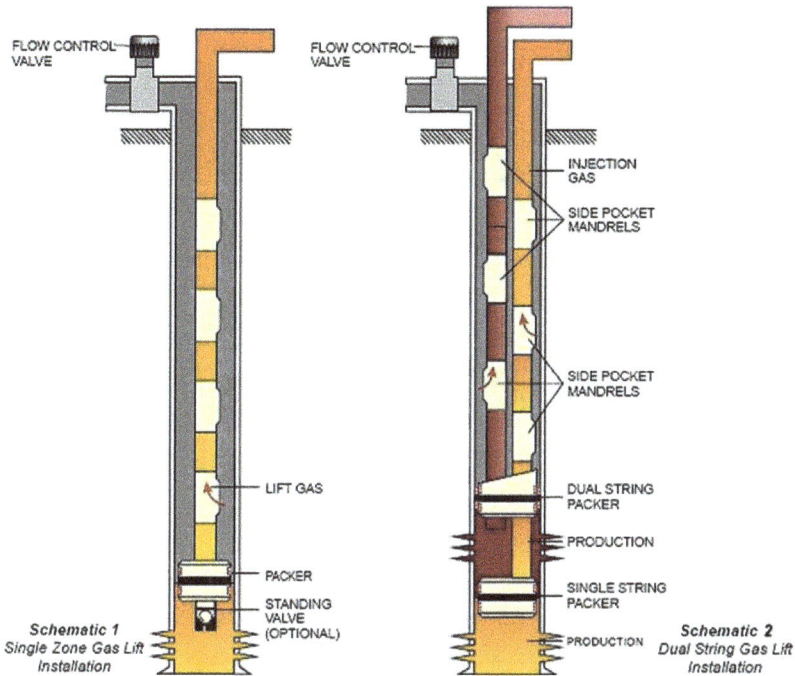

Figure 1.154. Gas lift arrangement.

Some of these roles are described below:

The **Company Drilling Supervisor** has overall responsibility for safety, efficient and effective drilling of the well in line with the well programme provided from the company's office. He will be seeking best value for money from Drilling Contractor and Service Companies. He is the most senior Company representative on the rig — hence sometimes called the *Company Man*. He is responsible for implementing the company's Safety Management System (SMS) on the rig. He manages the service companies, liaises with the company's office organisation, and ensures that optimum drilling parameters are being applied. He also reports the daily rig activities, manages logistics to and from the wellsite, and takes the lead in the event of well control problems. In most jurisdictions, the Company is liable for the safety and environmental compliance of the well itself. If the rig is close to other company operations (such as a jackup over an operating platform) the Company Man (or a more senior Company

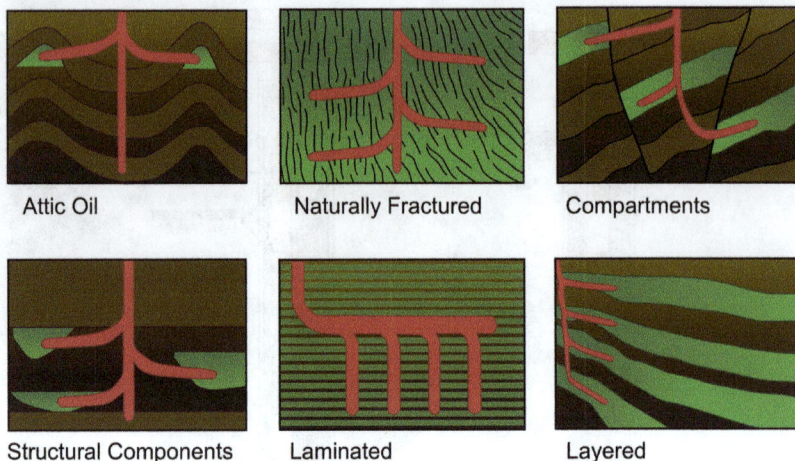

Attic Oil   Naturally Fractured   Compartments

Structural Components   Laminated   Layered

Figure 1.155. Multi-Lateral wells.

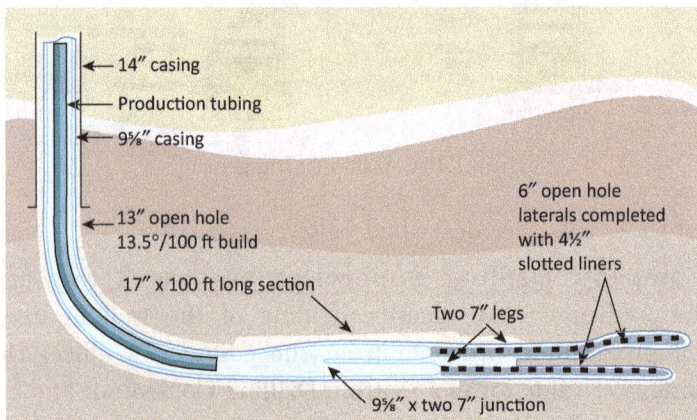

Figure 1.156. Multi-Lateral wells in the reservoir.

representative) will also be the "Offshore Installation Manager" or OIM. The OIM is legally responsible for the wellbeing of the entire combined operation.

The **Assistant Drilling Supervisor** (ADS) or **Wellsite Petroleum Engineer** (WPE) is sometimes a training role and reports to the Company Man. Their primary responsibilities are wellsite data collection and distribution, supervision of some service contractors (mud engineering, wireline logging and mud-logging), the

Figure 1.157. Multi-Lateral well to scale — Rio de Janeiro.

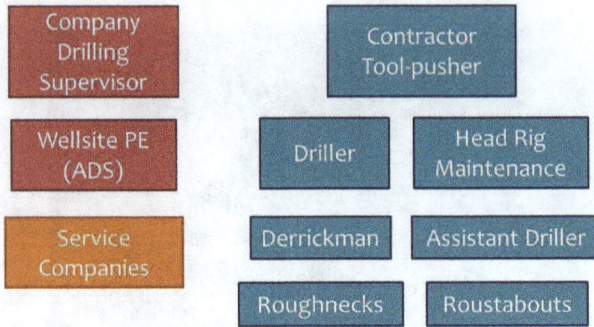

Figure 1.158. Rig organisation.

casing tally and records, cementing calculations and management of directional data.

There may well be additional Company staff on the wellsite depending on the on-going, activities such as **Safety** and/or **Compliance engineers**, Production Technologists (during production testing, Geologists (before and during coring), Petrophysicists (during logging), etc.

The **Contractor Toolpusher** is employed by the drilling contractor. He is accountable to the Rig Manager (an office-based role) and will be seeking to maximise the profit for the Drilling Contractor, which requires meeting the expectations of the Company Man while managing costs. He is responsible for the safe and efficient operation of the rig itself (as opposed to the well). He strives to avoid rig breakdowns and has an organisation that operates and maintains it. Where the rig is working alone, the Contractor Toolpusher is the OIM (Site Manager on land) and is responsible for the safety of all individuals on the wellsite.

For drilling operations, the **Driller** and the **Assistant Driller** operate the controls on the rig-floor, maintain the safety of the rig, maintain operating parameters provided by the Company Man and maintain well control. They are also responsible for the shift crews.

The **Derrickman** reports to the driller. His primary responsibilities are to work in the derrick managing the pipe (on the "monkey board"), managing the drilling fluid to parameters advised by the company, and mud pump maintenance.

Roughnecks (typically 3–5 on shift) report to the driller and generally work on the drill floor. They make pipe connections, nipple up/down BOPs and maintain equipment.

Roustabouts (typically 3–5 on shift) provide manual labour off the rig floor for moving equipment around on-site/on the well pad. All the above have roles in moving and rigging up a land-rig.

Rig Maintenance Supervisor is responsible for maintaining the drilling rig and support systems and the marine vessel in the case of mobile offshore mobile drilling units. He has a team that includes mechanical and electrical technicians, and Marine staff for managing the buoyancy and positioning of the rig.

The **Mud Engineer** works for a service company and reports at the wellsite to the ADS. He is responsible for the drilling/completion fluids. He manages the solids removal equipment, sampling and measurement of fluids, treatment of fluids and ordering additional chemicals as required.

The **Mud Logger** works for a service company and reports at the wellsite to the ADS. He runs a set of instruments that monitor mud volumes, flow, gas levels, etc, as well as taking and describing bit cuttings.

The **Casing Crew** work for a service company and report at the wellsite to the Company Man. They are responsible for running casing and tubing, applying the correct make-up torque and managing their equipment.

There are additional Service company staff that provide the following specialised roles, usually reporting to the Company

- cementing;
- petrophysical loggers;
- coring operations;
- running and operating MWD/LWD tools;
- directional drilling;
- fishing operations;
- jarring;
- waste management;
- navigation and positioning of the drilling rig (offshore);

- wellsite and road construction;
- running specialist completions equipment like ESPs, gauges;
- well testing;
- completion fluid filtering;
- fraccing;
- site security;
- helicopter operations;
- well control specialists;
- safety specialists for H_2S operation;
- maintenance of specialist drilling equipment TDS, pipe-handling;
- QA/QC of equipment, rig, etc;
- catering and cleaning of accommodation;
- certification inspectors;
- representatives from government authorities, etc;
- VIP visitors from Company, Rig Contractor, service companies, partners.

All in this team are required to work closely together to maximise safety and operational efficiency. The wellsite leadership on Company and Rig Contractor side are expected to demonstrate and promote a constructive working environment, where safety is taken very seriously.

Service contractors are typically paid for on a day or hourly basis and hence are called-off only when required, or as a contingency. There may also be rig accommodation limitations (especially offshore). Service contractors are responsible for the availability and correct operation of their own equipment. Overall coordination of service contractor activities on the wellsite is the responsibility of the Company Man.

Most positions on the wellsite work 12-hour shifts, so two crews are required. Some specialists work without shift coverage but must be rested (and if necessary operations shut down) as required. Crews also rotate on and off the wellsite; rotational arrangements vary widely but 28 days on, 28 days off is typical offshore or when crews need to travel long distances to the rig.

For the Company, office-based staff may be organised in roles as per Figure 1.159.

This well engineering team of course fits into a larger technical organisation comprising geologists, petrophysicists, production technologies, reservoir engineers, production engineers, etc. It is supported by commercial (contracting, procurement), logistics and HR/Finance capabilities.

1.34. Drilling Contracts, Procurement and Logistics

Given the need for specialist services, equipment and consumables, contractual arrangements need to be put in place. This is generally the responsibility of the Operating Company. Most contracts are essentially simple; equipment and manpower is called off on a day-rate basis. There may additionally be mobilisation/de-mobilisation lump-sum fees, and incentives for good and/or safe performance. A considerable proportion of contracts cover liabilities — who pays if there is an incident, damage or loss of equipment. A contract may run for a fixed period, e.g. one year, or a scope such as five wells.

Some contracts are on a lump sum basis (i.e. a single price irrespective of how long the operation takes) — including the drilling

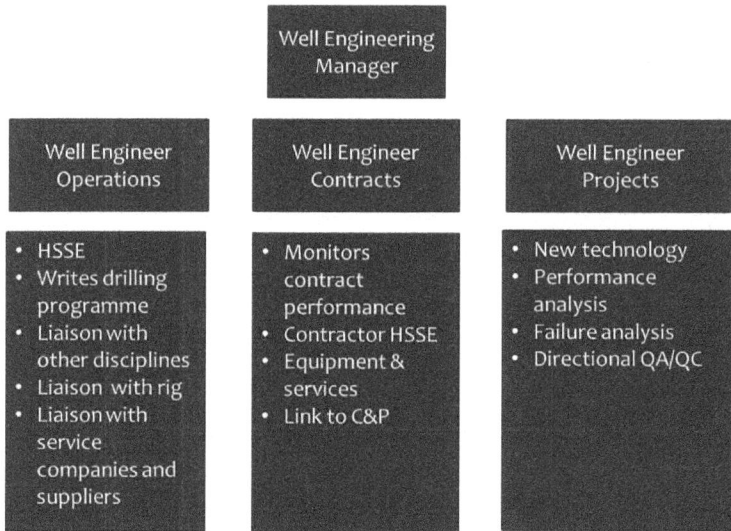

Figure 1.159. Operator's office organisation.

of entire wells in a few cases (called "Turnkey" drilling). If the scope is clear (e.g. there are no subsurface surprises), such contracts can work well to incentivise performance improvement and investment in new technology.

Procurement of equipment to be run in/on the well is based on a "functional specification" i.e. what the component needs to do (rather than prescribing an exact design). QA/QC and reliability are also important considerations, as well as availability/delivery dates and post-purchase support. Generally, as much as possible is pulled together into a single contract to obtain maximum leverage in the market.

Both equipment and service scopes are tendered to any party capable of meeting the requirements, which are identified before tendering. All else being equal, the tenderer with the lowest overall price is awarded the work and/or equipment order. Increasingly, previous performance is factored into the award decision.

Contracts are typically awarded for the following:

Services	Equipment/Consumables
• Drilling rig rental	• Wellhead
• Directional drilling/LWD/surveying	• Christmas tree
• Drilling fluids	• Casing
• Cementing	• Centralisers
• Petrophysical logging	• Tubing
• Mud logging	• Completion accessories
• Casing/tubing running	• Sand Screens
• Waste and cuttings management	• Gravel
• Other equipment specialists	• Artificial lift equipment
• Well testing	• Drilling fluid chemicals
• Inspection	• Cement and additives
• Logistics (boats, trucks, helicopters)	• Diesel
• Site construction	• Water
	• Lubricants
	• Personal Protective Equipment (PPE)

(Continued)

(*Continued*)	
Services	**Equipment/Consumables**
• Communications	• Rig and equipment spares
• Accommodation and catering (includes food)	
• Lifting	
• Environmental management	
• Local liaison	
• Security	

1.35. Logistics

Logistics covers a very wide range of activities, described in summary below.

1.35.1. *Land-Based Activities*

Logistics typically starts with **well pad construction**. This activity may range from simply clearing undergrowth to significant earth-moving to provide a flat location with suitable footing for a drilling rig (see Figure 1.7), and the building of a **cellar** for locating the wellhead. The cellar is a concrete-lined pit typically 2 m deep and 1 m × 1 m. The wellhead is set in the cellar to minimise drill floor elevation requirements when drilling and Christmas tree height once the well is completed. The well pad needs to be free from flooding risk, well-drained and designed to collect any spilt fluids. To minimise environmental impact, it should be no larger than necessary to support rig operations. Where necessary, steel or concrete plates are used to prevent the site deteriorating in wet weather.

Access to the well pad is needed to supply the rig — usually by road. In many locations, **roads** will need to be built specially for the drilling operation — these will usually be left in place after operations are complete, if requested by local residents. Roads need to be sized for rig transportation. In desert locations, roads require very regular maintenance to prevent severe degradation.

All the above will require land access, permits, local government approvals and often an environmental risk assessment to ensure minimum disruption to the environment. Getting these agreements in place — typically alongside similar activities for production facilities and pipelines — can take some time and demands proper project planning.

Supplying the rig will require attention to **warehousing** and **transport**, usually by truck, to ensure that the rig is never kept waiting on equipment or materials. Often, suppliers and contractors are responsible for their own logistics; sometimes, the operator is responsible for this. Transport to and from the rig is a significant source of safety risk.

1.35.2. *Offshore Activities*

Offshore, the focus is on shipping and helicopter operations. **Shipping** is usually by supply boat to and from a rig. Each operator usually contracts his own supply boats, which are used to transport all party's requirements to and from the rig. Again, it is essential that the rig is not left waiting on equipment or supplies (overall operations costs can reach $1 million/day). Supply boats can cost $50,000 per day.

Helicopters are used to transport all crew to and from the rig, along with small items of equipment — there are limitations on size, weight and content. Explosives or concentrated chemicals require special clearance. Rig crew require special training on helicopter evacuation in the event of a ditching, and in cold climates wear **survival suits** that insulate against the cold, should such an event occur. Helicopters are contracted by the operator and can cost $3,000 per flying hour. Helicopters are also normally used in the event of rig evacuation for safety reasons.

A **standby boat** is typically kept stationed within 500 m of the drilling rig. If lifeboats are deployed from the rig, the standby boat will support recovery activities, and it is also available should anyone fall overboard.

In all rig operations, it is essential that inventory in **warehouses** and in transit is actively managed and the location of items is tracked so that location is known at all times. Computer systems typically manage this — reordering, for example, when stocks get low. In more sophisticated locations, vehicles are tracked and monitored by

GPS tracking systems as part of **Journey Management** and smart (RFID) tags are used on items of equipment.

1.36. Performance Improvement

The very significant cost of drilling and completion activities invites seeking continuous improvement. Generally, this will impact cost, duration and safety of activities. The most significant area of progress is in the development and application of new technology. The following technologies have been developed and widely applied over the last 30 years in this business:

- horizontal wells;
- extended reach wells;
- geo-steering;
- unbalanced drilling;
- MWD/LWD;
- automated/dual-activity rigs;
- subsea wells;
- expandable tubulars;
- wired drill pipe;
- swellable elastomers;
- probabilistic casing design;
- downhole sensors;
- rig instrumentation;
- reliable ESPs;
- metal-to-metal wellhead seals;
- well design software;
- PDC bits;
- many new logging tools;
- low-toxicity oil-based mud;
- casing threads/Joint Makeup Analysis (JAM).

This innovation has come mainly from operating and service companies, in addition to small entrepreneurial organisations. However, the industry remains risk-averse and trialling a new technology can be resisted by those in the field.

There have also been experiments in different contracting strategies in an attempt to align objectives of company, rig contractor and service companies. For example, in the 1990s, the focus was on outsourcing as much as possible to the rig contractor. Experience showed that drilling contractor companies at that time did not possess the skills or motivation to take on this expanded role, and operating companies have since taken back these responsibilities. More successful has been the development of Integrated Service Companies (ISCs) such as Baker Hughes, Halliburton and Schlumberger, who acquired smaller companies and who can now provide a large proportion of services under a single contract.

Performance has also been raised by following a structured Well Delivery Process (WDP) as described at the start of this chapter. By dissecting activities into small steps, targets can be set and improvements made that can make a significant overall difference. Technical Limit Drilling (TLD) is one such process that originated in Woodside Petroleum, and was subsequently developed further and applied in various guises in Shell, BP, Exxon, Chevron, BG Group and other operating companies.

Taking a simple example, the well may require the following basic steps, each of which will be challenged to produce target duration. These performance improvements are often shown on a Time Versus Depth (TVD) plot as shown in Figure 1.160.

Activity	Depth	Duration	Target
Spud	0	0	0
Drill 36″ hole	200	2	1
Run 30″ conductor	200	2	1
Drill 26″ hole	1,600	2	1
Run & Cement 20″ casing	1,600	5	3
Drill 17$1/2$″ hole	3,000	2	1
Run & Cement 13$3/8$″ casing	3,000	4	3
Drill 12$1/4$″ hole	5,700	4	3
Run & Cement 9$5/8$″ casing	5,700	4	2
Drill 8$1/2$″ hole	11,000	10	8
Run & Cement 7″ casing	11,000	5	3
Run completion	11,000	5	4

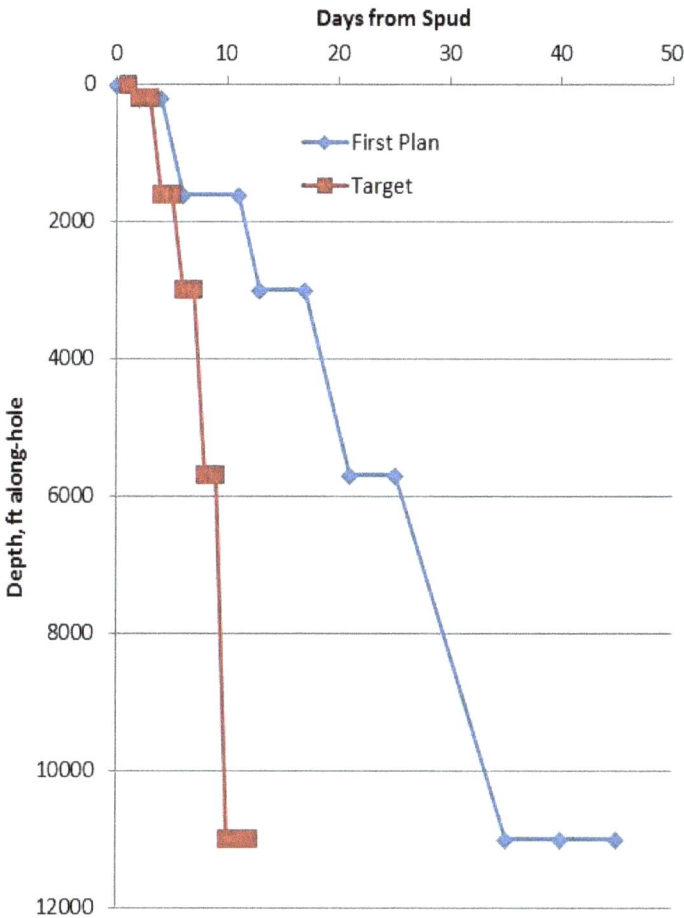

Figure 1.160. Typical Time Versus Depth (TVD) plot.

1.37. Example Daily Operations Report

Figure 1.161 shows an example of a **Daily Operations Report (DOR)** for a land drilling operation.

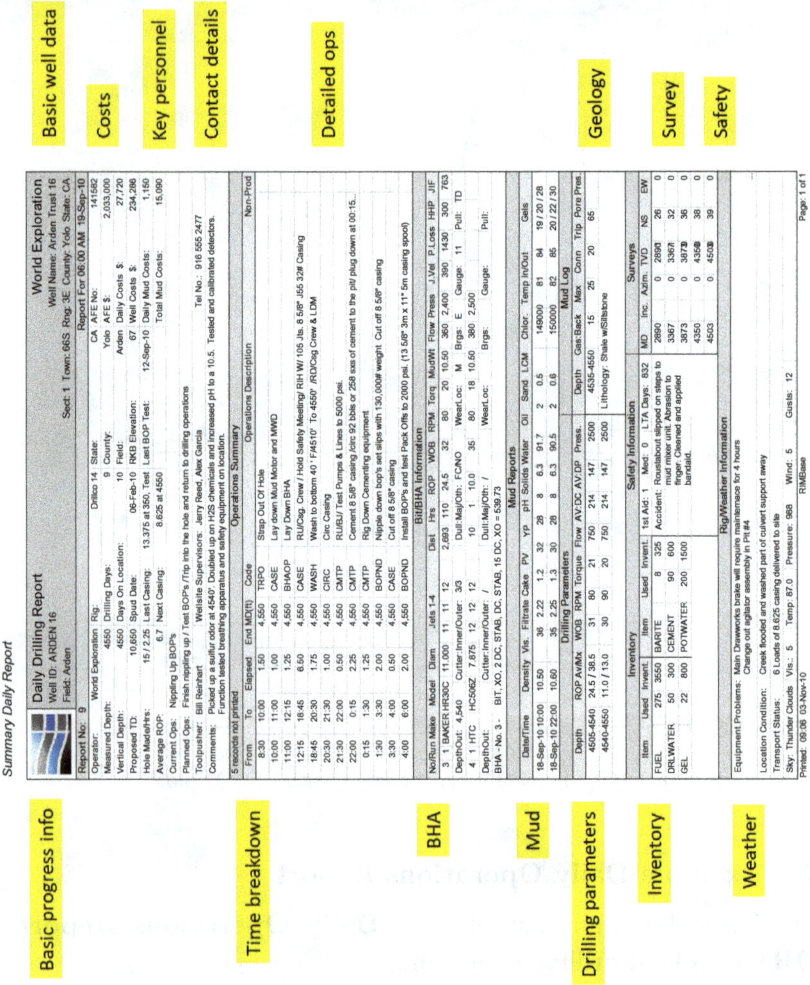

Figure 1.161.   Typical daily operations report.

1.38. Glossary of Terms

Term	Meaning
ADS	Assistant Drilling Supervisor
API	American Petroleum Institute
BHA	Bottom Hole Assembly
BOP	Blow Out Preventer
CAPEX	Capital Expenditure
CBL	Cement Bond Log
CO_2	Carbon Dioxide
CRA	Corrosion Resistant Alloy
CTD	Coiled Tubing Drilling
DC	Drill Collar
DCF	Discounted Cash Flow
DOR	Daily Operations Report
DP	Drill Pipe
DP	Dynamic Positioning
DSV	Drilling Supervisor
ECD	Equivalent Circulating Density
EMV	Estimated Monetary Value
EOB	End Of Build
ESP	Electrical Submersible Pump
EU	External Upset
FPSDO	Floating Production Storage Drilling and Offloading
FPSO	Floating Production Storage and Offloading
GMS	Gyro Multi Shot
GoM	Gulf of Mexico
GOC	Gas Oil Contact
GPS	Global Positioning System
GWC	Gas Water Contact
H_2S	Hydrogen Sulphide (poisonous gas)

(Continued)

(*Continued*)

Term	Meaning
HSE	Health, Safety and Environment (considerations in operations)
HWDP	Heavy-Weight Drill Pipe
ICV	Inflow Control Valve
ID	Inside Diameter
JAM	Joint Analysed Makeup
KOP	Kick-off Point
KPI	Key Performance Indicator
LCM	Lost Circulation Material
LWD	Logging While Drilling
MD	Measured (along hole) Depth
MMS	Magnetic Multi Shot
MSS	Magnetic Single Shot
MTBF	Mean Time Between Failures
MWD	Measurement While Drilling
OBM	Oil Based Mud
OD	Outside Diameter
OOC	Oil On Cuttings
OPEX	Operations Expenditure
OWC	Oil Water Contact
PCP	Progressive Cavity Pump
PPE	Personal Protective Equipment
RFID	Radio-Frequency Identification (SMART tags on equipment)
RPM	Revolutions Per Minute
RSS	Rotary Steerable System
SAGD	Steam Assisted Gravity Drainage
SBM	Synthetic Oil-Based Mud
SPM	Side Pocket Mandrel

(*Continued*)

(Continued)

Term	Meaning
SSD	Sliding Side Door
SSSV	Subsurface Safety Valve
SSTT	Sub Surface Test Tree
TCP	Tubing Conveyed Perforating
TD	Total Depth
TDT	Thermal Decay Time (logging tool)
TDS	Top Drive System
TLC	Tough Logging Conditions
TLD	Technical Limit Drilling
TOC	Top of Cement
TVD	Time Versus Depth (plot)
UBD	Under Balanced Drilling
WBM	Water Based Mud
WDP	Well Delivery Process
WOB	Weight on Bit
WPE	Wellsite Petroleum Engineer
QA/QC	Quality Assurance/Quality Control

Acknowledgements

The author wishes to acknowledge those at Shell, BG Group, Imperial College, in other organisations, and family and friends for their help and support in writing this chapter. The author will be donating all royalty fees from this publication to The John S. Archer Endowment Fund — http://www.imperial.ac.uk/earth-science/research/research-groups/perm/events/john-archer-endowment/. Every effort has been made to trace the copyright holders and obtain permission to reproduce this material.

Figure #	Acknowledgement
1.1	James Reason
1.5	Wikipedia (public domain)
1.6	Image Credit: Greg Bright - Ocean Fab
1.7	Courtesy of Abdulharim Md Gamal, Egypt
1.9	Major Drilling Group International Inc.
1.10	Avalon Licensing Ltd
1.11	Courtesy of Dr Horst Kreuter
1.12	Alamy
1.14	Offshore Energy/Heerema
1.15	designersparty.com
1.16	Kvaerner
1.17	Kvaerner
1.18	Getty
1.19	Shell International Limited
1.20	Marine Traffic
1.21	Maersk Drilling A/S
1.22	Finn Tornquist
1.23	Dock Wise
1.24	Seadrill
1.25	National Oilwell Varco (NOV)
1.26	FlowServe
1.27	SeaDrill
1.28	OSHA
1.29	Savannah Energy Services Corp.
1.30	Metroforensics
1.32a	National Oilwell Varco (NOV)
1.32b	National Oilwell Varco (NOV)
1.32c	OSHA
1.32d	National Oilwell Varco (NOV)
1.32e	Steve Devereux

(*Continued*)

(*Continued*)

Figure #	Acknowledgement
1.33	NIOSH (public domain)
1.35	Deepak Choudhary
1.36	National Oilwell Varco (NOV)
1.37	osha.gov ©University of Texas at Austin
1.38	National Oilwell Varco (NOV)
1.40	National Oilwell Varco (NOV)
1.41	National Oilwell Varco (NOV)
1.42	drillingformulas.com
1.43	drillingformulas.com
1.44	Guzelian
1.45	World Oil
1.46	Roy Luck
1.47	OSHA
1.48	Central Mine Equipment Company
1.50	CleanTech
1.52	Vallourec
1.53	Kevin Copple
1.54	National Oilwell Varco (NOV)
1.56	TIX Holdings Company Limited
1.59	Stavanger Oil Museum
1.60	Baker Hughes
1.61	Baker Hughes
1.62	TIX Holdings Company Limited
1.63	Baker Hughes
1.68	Schlumberger Limited (SL) — Image and any associated trademarks owned by SL
1.69	Enventure
1.71	Schlumberger Limited (SL) — Image and any associated trademarks owned by SL

(*Continued*)

(*Continued*)

Figure #	Acknowledgement
1.72	Mono-block wellhead
1.73	OneSubsea
1.74	Schlumberger Limited (SL) — Image and any associated trademarks owned by SL
1.75	HartmannValves (https://commons.wikimedia.org/wiki/File:Wellhead_Bohrlochkopf.JPG), "Wellhead Bohrlochkopf", https://creativecommons.org/licenses/by/3.0/legalcode
1.76	Alamy / Shell
1.77	Saab Seaeye Ltd.
1.78	Schlumberger
1.79	Schlumberger Limited (SL) — Image and any associated trademarks owned by SL
1.80	Schlumberger Limited (SL) — Image and any associated trademarks owned by SL
1.81	Ovo Egguono Nigeria Ltd
1.82	Schlumberger Limited (SL) — Image and any associated trademarks owned by SL
1.83	Halliburton
1.84	Schlumberger Limited (SL) — Image and any associated trademarks owned by SL
1.85	Halliburton
1.86	vautron.com.au
1.88	Baker Hughes
1.89	Halliburton
1.90	Schlumberger Limited (SL) — Image and any associated trademarks owned by SL
1.91	Forum Energy Technologies, Inc.

(*Continued*)

(Continued)

Figure #	Acknowledgement
1.92	Schlumberger Limited (SL) — Image and any associated trademarks owned by SL
1.93	Halliburton
1.95	Centek Limited
1.99	Prof. Pidwirny, Michael, University of British Columbia
1.105	Baker Hughes
1.106	Halliburton
1.108	Baker Hughes
1.109	National Energy Board of Canada
1.111	Shell International
1.112	Shell International
1.115	Halliburton
1.116	Schlumberger Oilfield Review, Summer 1999 [see Note 1 below]
1.117	Schlumberger Oilfield Review, Summer 1999 [see Note 1 below]
1.118	Schlumberger Oilfield Review, Summer 1999 [see Note 1 below]
1.119	Schlumberger Oilfield Review, Summer 1999 [see Note 1 below]
1.120	Schlumberger Oilfield Review, Summer 1999 [see Note 1 below]
1.121	Schlumberger Oilfield Review, Summer 1999 [see Note 1 below]
1.122	Schlumberger Oilfield Review, Summer 1999 [see Note 1 below]
1.123	Schlumberger Oilfield Review, Summer 1999 [see Note 1 below]
1.125	Schlumberger Oilfield Review, Summer 1999 [see Note 1 below]

(Continued)

(*Continued*)

Figure #	Acknowledgement
1.126	Schlumberger Oilfield Review, Summer 1999 [see Note 1 below]
1.127	Schlumberger Oilfield Review, Summer 1999 [see Note 1 below]
1.128	Schlumberger Oilfield Review, Summer 1999 [see Note 1 below]
1.129	Schlumberger Oilfield Review, Summer 1999 [see Note 1 below]
1.130	Halliburton
1.131	Halliburton
1.132	Wikipedia user Blastcube (https://commons.wiki media.org/wiki/File:Diamond_Core.jpeg), "Diamond Core", https://creativecommons.org/licenses/by-sa/3.0/legalcode
1.133	Photo: Mark Pomeroy
1.134	International Ocean Discovery Program [See Note 2 below]
1.136	Crain's Petrophysical Handbook
1.138	American Oil & Gas Historical Society
1.140	hvmag.com/Al Granberg/Propublica
1.141	Dave Yoxtheimer, Penn State Marcellus Center for Outreach and Research
1.142	Owner: Texas Energy Museum
1.143	University of California Berkeley/Deepwater Horizon Study Group
1.144	Helix Energy Solutions Group, Inc
1.145	Source: Oil & Gas UK
1.146	Ian West Geology Photographs

(*Continued*)

(*Continued*)

Figure #	Acknowledgement
1.147	TastyCakes on English Wikipedia (https://commons. wikimedia.org/wiki/File:Pump_Jack_labelled.png), "Pump Jack labelled", https://creativecommons.org/ licenses/by/3.0/legalcode
1.148	Image courtesy of Octal Steel
1.150	Baker Hughes
1.151	Baker Hughes
1.152	Schlumberger Limited (SL) — Image and any associated trademarks owned by SL
1.153	Baker Hughes
1.154	American Completion Tools
1.157	Statoil
1.161	Infostat Systems Inc.

Note 1: Figures 1.116–1.123 and 1.125–1.129 are illustration copyright Oilfield Review, used with permission of Schlumberger. The author acknowledges Aldred W, Plumb D, Bradford I, Cook J, Gholkar V, Cousins L, Minton R, Fuller J, Goraya S and Tucker D. Managing drilling risk, *Oilfield Review* 11, no. 2 (Summer 1999): pp. 2–19.

Note 2: Figure 1.134: Expedition 319 Scientists, 2010. Methods. In Saffer D, McNeill L, Byrne T, Araki E, Toczko S, Eguchi N, Takahashi K, and the Expedition 319 Scientists, Proc. IODP, 319: Tokyo (Integrated Ocean Drilling Program Management International, Inc.). doi:10.2204/iodp.proc.319.102.2010.

Chapter 2

Core Analysis

Vural Sander Suicmez[*,¶], Marcel Polikar[†,¶], Xudong Jing[‡]
and Christopher Pentland[§,¶]

[*]Maersk Oil & Gas A/S, Copenhagen, Denmark
[†]Consultant, Montreal area, Canada
[‡]Shell Global Solutions International BV, The Netherlands
[§]Petroleum Development Oman, Muscat, Sultanate of Oman

2.1. Introduction

The term *petrophysics* was first introduced by Gus E. Archie [1], a Shell engineer, in the context of laboratory measurements of core material from petroleum reservoirs. Although the concept has expanded to encompass well log data acquisition and interpretation, core analysis remains an important element within the domain of petrophysics.

Core analysis can be defined as the laboratory measurement of the physico-chemical properties of samples of recovered core, for purposes within multiple disciplines. A geologist, for example, needs core analysis for facies analysis, mineral identification or clay typing, or to obtain depositional information and build static reservoir models. A reservoir engineer uses core analysis for comprehensive interpretation of fluid flow characteristics in field applications in order to design and optimise the recovery processes. A production technologist would obtain information on well injectivity, sand

---

[¶]Formerly with Shell Global Solutions International BV, The Netherlands.

control parameters, rock mechanics parameters for fracture design and mineralogy for acid stimulation. A petrophysicist in general not only has the task of the organisation and implementation of the coring and core-analysis programme but also designs core-analysis measurements for the purpose of calibrating logs and determining input parameters into log-interpretation models. The wide variety of information that can be gathered from recovered core samples requires the involvement of specialists from different sub-surface technical teams. Figure 2.1 summarises the requirements of different disciplines from core analysis.

It is clear that economic and efficient development of hydro-carbon reservoirs is highly dependent on understanding the key properties of reservoirs, such as porosity or permeability, and rock–fluid interactions such as wettability. Furthermore, solving fluid-flow equations in porous media requires functions of multi-phase transport properties, such as capillary pressure and relative permeability. These data can only be gathered through a carefully conducted core-analysis program. Core samples are considered to be one of the most direct and valuable sources of data for sub-surface studies [2].

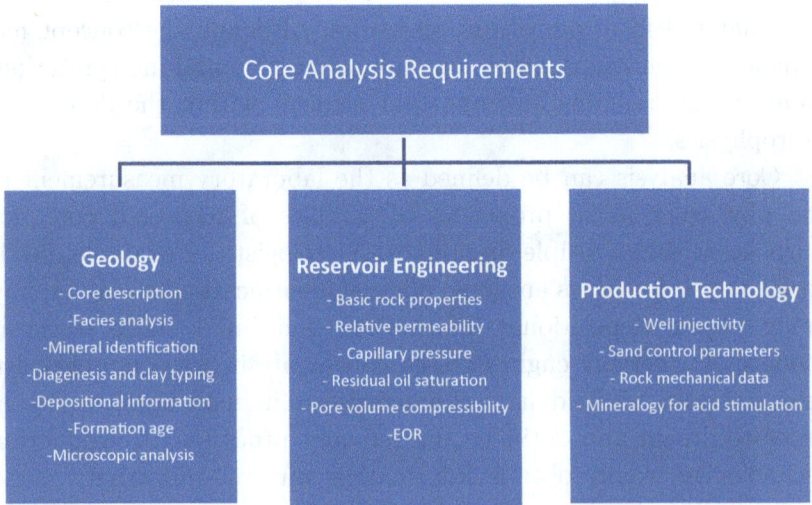

Figure 2.1. Core analysis requirements for different technical disciplines.

Before going into the details of laboratory measurement techniques, we discuss the process of core sampling, handling and preservation.

2.2. Formation Samples

Formation samples are divided into two subgroups: (1) drill cuttings and (2) core samples (see Figure 2.2).

2.2.1. *Drill Cuttings*

Drill cuttings are circulated by drilling mud and are collected at the surface for examination. They provide useful information on penetrated layers in the sub-surface, such as mineralogical composition, pore-size distribution of the rock and possible fluid content of the layers.

Although they provide timely and relatively inexpensive data, the reliability and accuracy of information derived from drill cuttings are limited. One of the uncertainties is associated with the depth. The

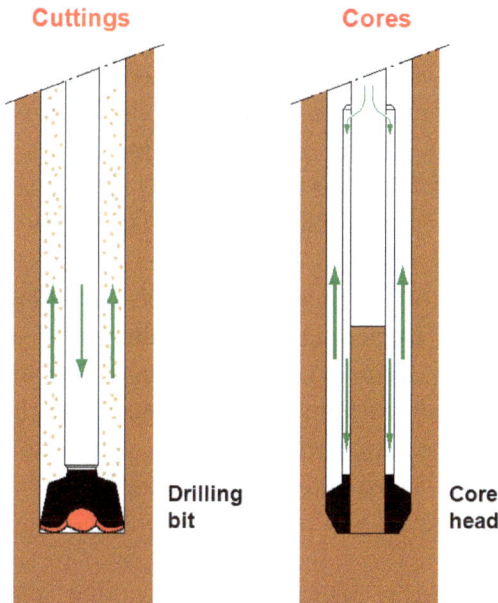

Figure 2.2. A schematic showing formation sampling. Drill cuttings (left) and core sample (right).

cuttings travel to the surface with the mud, a journey which may take a few hours as the travel time depends upon the friction of the cuttings in the mud and the circulation hydraulics; therefore, depth of formation is often not very accurately defined.

Another shortcoming is associated with the size and irregular shapes of the cuttings. As the cuttings are too small and have been disturbed by the drill bit, accurate quantitative petrophysical measurements like porosity and permeability may not be possible. In addition, the cuttings are often contaminated by the drilling mud. However, the information obtained through cuttings can be complementary to that produced by the more accurate and sophisticated evaluation methods of core plug analysis.

2.2.2. *Coring*

Coring is the process of cutting and removal of rock samples from the wellbore. Core samples can be considered to be the most direct source of information on sub-surface rocks; however, if the coring process is not carefully designed, it may be impossible to obtain good quality data. It is worthwhile to understand the impact that the coring process can have on a core-analysis programme as a whole. Events that happen during coring can later play a critical role in the interpretation of the measurement results. From the moment the drill bit comes close to the core material, a variety of alteration processes commence [3]. During the normal drilling process with a slight overpressure, the formation is under a greater pressure from the mud column in the well than from the fluid in the formation. The differential pressure across the wellface causes mud and mud filtrate to invade the formation immediately adjacent to the well surface, thus flushing the formation with mud and its filtrate. When drilling is done with water-based mud, water filtrate invades the core and displaces some of the oil and perhaps some of the original interstitial water [4]. Therefore, the core may get exposed to the drilling fluid and cause alteration of physical properties that affect the desired measurements, such as permeability or porosity. Moreover, while the sample is being brought to the surface, the confining pressure of the fluid column is constantly decreasing. This reduction permits the expansion of the entrapped water, oil and gas.

Gas, having a greater coefficient of expansion, expels oil and water from the core. Thus, the fluid contents of the core at the surface can be significantly different than those in the formation [4]. Therefore, extra care should be taken while planning and executing a coring programme. Such considerations include the type of bit, type of core, drilling mud composition, length and type of core barrel, and the rate at which core should be brought to surface [5]. Proper considerations maximise the amount of core recovered. Furthermore, the case for coring requires a clear definition of the objectives, and in many cases economic justification. The added value of a coring programme in terms of reducing uncertainty in rock and fluid properties should be clearly indicated. Only after a clear need for a core-analysis programme is satisfactorily justified, and the necessary technical precautions are taken, should a coring programme be initiated.

There are two main types of coring: (1) coring axial to the well bore which is called bottomhole continuous coring and (2) coring from sidewall of the borehole which is called sidewall coring.

2.2.2.1. *Bottomhole continuous coring*

As the name suggests, this method cuts the core at the bottom of the borehole, axial to the wellbore. Bottomhole continuous cores are obtained by using a core barrel attached to the bottom of the drill pipe. A coring bit is attached to the outer barrel and a core catcher is fitted to the bottom of the inner barrel. A schematic of the core barrel and coring bit is shown in Figure 2.3. A cylindrical piece of formation, which is often 9 m or 18 m in length and 10 cm in diameter, is put into the core barrel and taken to the surface. Although it is possible to cut the core at different lengths and diameters, a tendency to reduce the core diameter leads to faster and hence cheaper coring operations. It is also possible to cut oriented cores, which have lines scribed along their length to show the core position relative to the core head. The orientation of the core head is measured by similar techniques to those used to measure the deviation of a well. Oriented cores are particularly valuable where fractures exist in the sub-surface.

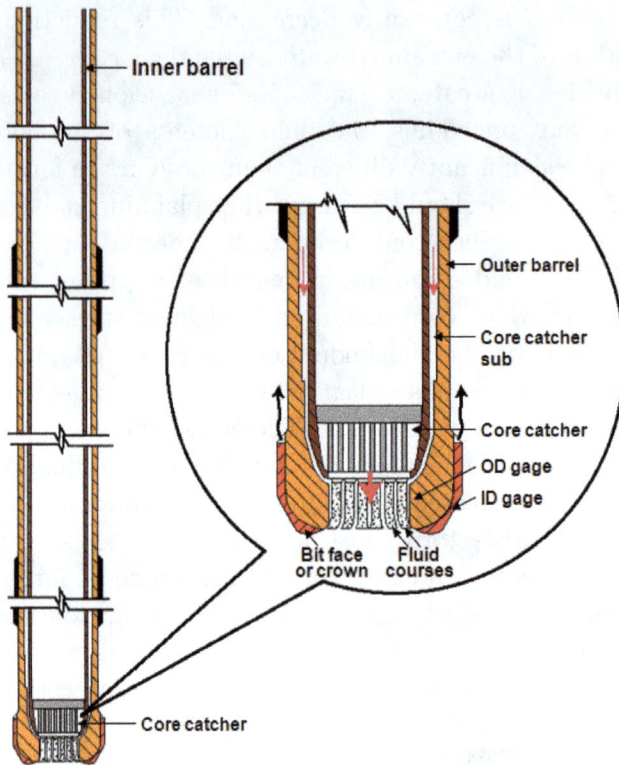

Figure 2.3. A schematic showing the core barrel and the bit attached to the bottom of the drill pipe.

Standard coring techniques have been utilised with varying degrees of success as they suffer from certain limitations. Damaging the core material (especially for unconsolidated formations), mud invasion, and the high costs associated with some coring operations are the major issues confronting the core-analysis industry. Some of these issues can be addressed by several specialised coring techniques, as each provides specific benefits [6]. Sponge coring, gel coring and Coring-While-Drilling (CWD) are examples of specialised coring techniques.

Sponge coring was developed to improve the accuracy of core-based oil saturation [7]. The core fluids contained in the retrieved core are expelled as a result of depressurisation as the core is

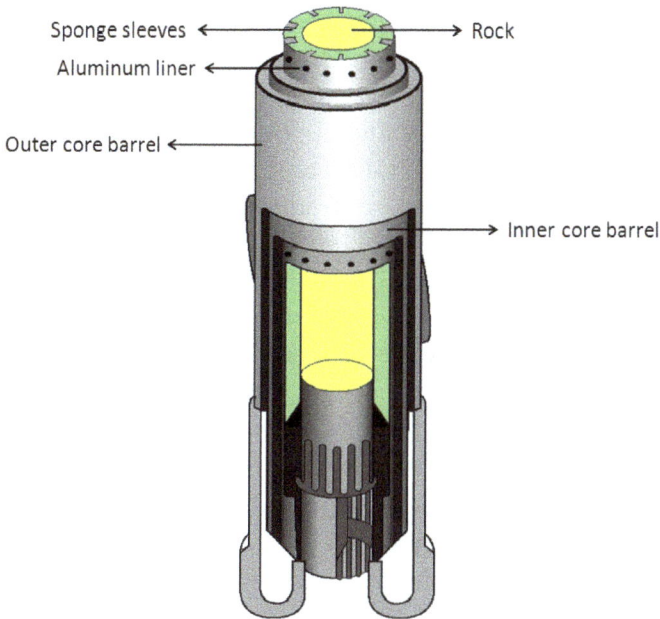

Figure 2.4. A schematic showing a sponge core barrel (Courtesy Security DBS).

brought to the surface, but are collected and trapped in an absorbent polyurethane material (sponge) lining the inner barrel. The sponge is composed of an open-celled foam with a porosity of 70–80% and a very high interconnectivity between the cells (see Figure 2.4). The sponge can absorb and collect a volume of fluid up to one order of magnitude larger than the fluid capacity of most rock materials. Depending on the core analysis objectives, the sponge can be made preferentially oil- or water-wet. An oil-retentive sponge is used with water-based mud. A water-retentive sponge is used with oil-based mud.

An alternative tool, which is called liquid trapper, was recently introduced by Corpro Group to allow accurate and direct saturation measurement from coring. The inner barrel of the liquid trapper is equipped with an inflatable-seal system, which traps the liquids escaping from the core during the retrieval to the surface. A dual-seal system, which consists of an upper and a lower seal, separates the core inside the inner barrel into 1 m (3 ft) core/fluid closed compartments.

At the wellsite, the fluid draining from the compartments is immediately collected, and hence the total volume of the expelled liquid is estimated. The remaining oil saturation can then be calculated by adding the volume of oil obtained from the liquid trapper to the oil volume measured in the adjacent core. A more sophisticated coring tool, QuickCapture, was later introduced by the same group. The system is designed to capture 100% of gas on top of the liquids released from the core. One of the key innovative features of this technique is the depressurisation and sampling of all gases and liquids in the downhole rather than at the surface. A sealed assembly capped at 500 psi brings the core to the surface, while retaining all gases and liquids which are expelled from the core during the trip out.

Gel coring uses high viscosity gel for downhole core encapsulation and preservation, which is an alternative to operator-intensive wellsite core preservation. Core gel is a viscous, high molecular weight, polypropylene glycol with zero spurt loss, which is non-soluble in water and environmentally safe. Because the gel comes in direct contact with the core during and immediately after it is cut, further exposure to core contaminants is minimised. The high viscosity gel stabilises poorly consolidated rocks with moderate compressive strengths and enhances core integrity. Core gels can be customised to address a wide range of coring situations and rock types [6].

The CWD system is designed to provide operators with the flexibility of bottomhole coring or drilling with the same bit, without tripping-out of the borehole. In the drilling mode, the system is used in the same manner as a conventional bottomhole assembly. In the coring mode, a drill bit plug is replaced with an inner barrel and bearing assembly that transforms the drill bit into a core bit. After core recovery, the coring assembly is retrieved with a wireline and overshot assembly. A CWD system significantly reduces the time and therefore the cost necessary to cut continuous full-diameter cores [6].

2.2.2.2. *Sidewall coring*

Sidewall sampling is another means of obtaining reservoir rock samples, if borehole conditions do not allow full-diameter continuous

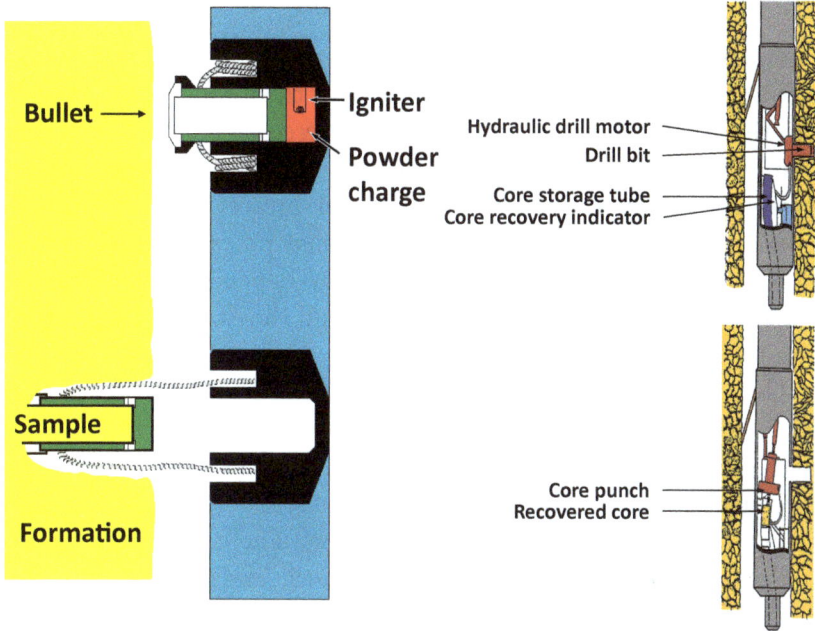

Figure 2.5. Schematics of a sidewall sample gun (left), which is used to obtain percussion sidewall cores, and a rotary sidewall-coring tool (right).

coring. It is also cheaper than continuous axial coring. Rock samples are obtained either by firing hollow cylindrical bullets into the borehole wall, which is called percussion sidewall coring, or by drilling a small horizontal core in the same way as plugs are cut from full-diameter core, which is called rotary sidewall coring [2]. Figure 2.5 shows the schematics of a sidewall sample gun, which is used to obtain percussion sidewall cores, and a rotary sidewall-coring tool. The advantages of this technique are that obtaining sidewall core material is quick and relatively inexpensive, the exact depth of coring is known, and recovered samples are much larger than drill cuttings, which enables better evaluation of geological variations and quantitative petrophysical analysis [2]. A disadvantage of sidewall coring is that samples (especially percussion cores) are often damaged and hence they may be unsuitable for laboratory tests. Moreover, sample volumes are usually insufficient for performing advanced studies, such as multi-phase flow and relative permeability measurements.

2.3. Core Handling and Preservation

The main objective of a coring and core-preservation programme is to obtain the rock that is representative of the formation and deliver it to the core analysis laboratory as unaltered as possible. Although a variety of techniques have been developed, there is unfortunately no single best method for handling and preserving core samples. Different rock types may require additional precautions. The most appropriate core handling techniques may depend on the length of time for transportation, type of storage, and the nature of the specific tests to be conducted [8].

Conventional core handling involves breaking lengths of core into 1 m ($\sim$3 ft) sections as the core is retrieved from the inner barrel on the drill floor or on the catwalk. These pieces are then loaded in sequential order into transit core boxes and taken to the place designated for core layout and description. The core should be laid out, cleaned, fitted, marked and described, and then packed into the final transport boxes for shipment to the core storage/analysis facilities [9].

Containerised cores, however, need certain additional precautions to prevent or limit core disturbance. Flexure of inner barrels has been shown to induce significant core damage through cracking. This problem is especially marked for the smaller diameter barrels and is more pronounced for fibreglass and plastic liners than it is for other more rigid materials. In order to limit this disturbance, it is essential that the inner barrels/liners be supported with rigid structural supports. Core cradles designed to hold 9 m ($\sim$30 ft) lengths of inner barrel are considered to be the optimal solution for this purpose in terms of preventing flexure while safely lifting the core. Space restrictions (small or enclosed drill floors) may preclude the use of such cradles. The cradles are lifted to the catwalk or other area where the barrel is lifted out of the cradle and laid down. The core can then be processed (e.g. measuring, plugging, cutting, sampling, repacking) according to the previously agreed programme [9].

Unconsolidated cores normally require some form of stabilisation inside the core barrels prior to transport. Two stabilisation methods are in general use in the industry: freezing of the core and injection of

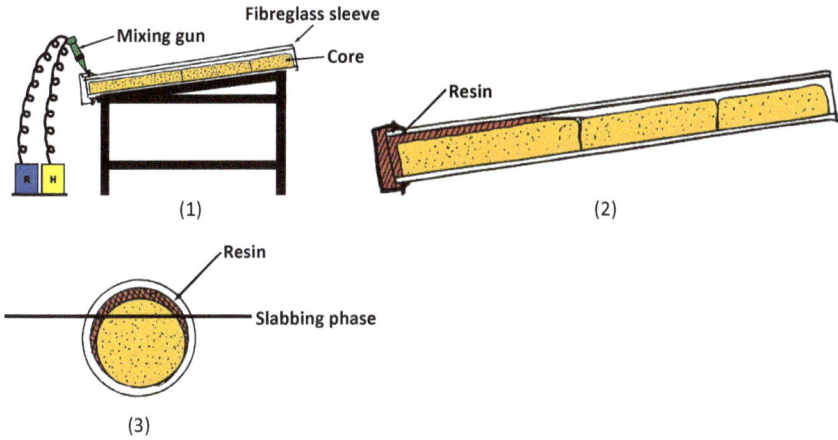

Figure 2.6. An overview of a core-handling programme for unconsolidated core samples: (1) injection of fast-hardening plastic (resin), (2) resin fills the core and annulus, and (3) slabbing takes place.

fast hardening epoxy resin/plastics in the core/core barrel annulus. Often a combination of both methods is used [9]. Figure 2.6 gives an overview of core-handling procedures for unconsolidated rocks.

Sponge cores may require precautions similar to those applicable to other containerised cores, although special procedures are required for the preparation of the core barrels prior to coring, and for the packing and sealing of the core barrels after retrieval [9]. Because of the specialised nature of this type of coring, requirements should be discussed in detail with the contractor involved and with the end-users of the data.

2.4. Core Analysis Preparation

After the core has been cut and preserved at the wellsite, it is transported to the core analysis laboratory where it is subjected to a wide variety of measurements. These measurements are divided into two main categories: (1) basic (or Routine) Core Analysis Laboratory (RCAL) and (2) Special Core Analysis Laboratory (SCAL) measurements. They include grain density, porosity, permeability, fluid saturation, electrical resistivity, capillary pressure and relative permeability measurements. However, several steps should be taken

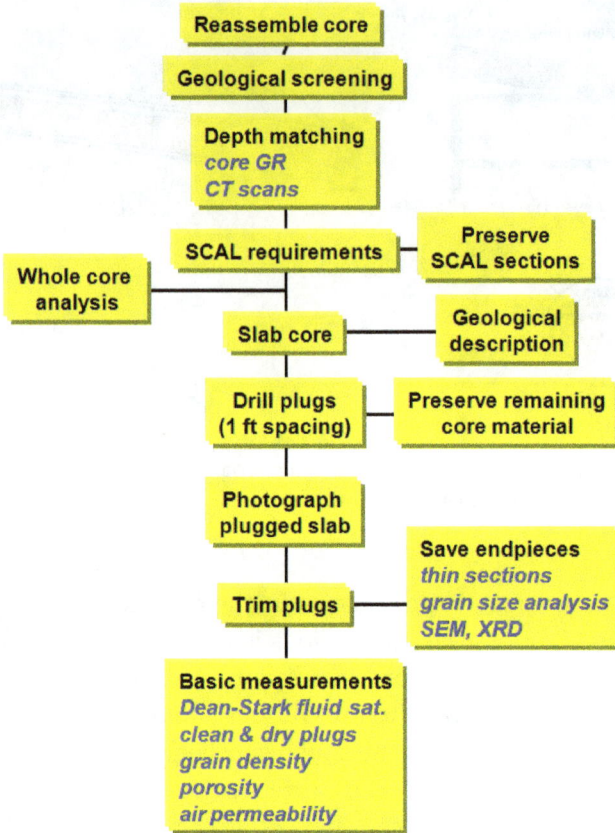

Figure 2.7. Recommended flow diagram for basic core analysis [5].

before these measurements take place. These are imaging, sample selection, core plugging and plug preparation. The flow diagram for conducting basic core-analysis measurements is shown in Figure 2.7.

2.4.1. *Core Description and Imaging*

Once the core is retrieved, gamma ray logging is performed as the first step towards core analysis. The core gamma ray logger is a portable device that provides gamma ray logs immediately after the core has been cut and retrieved. This analysis can be conducted either at the wellsite or in the laboratory. The main purpose of it is to correlate the depth of the cored sections with the anticipated lithology by delineating the shale sections from the non-shale (reservoir) intervals. Reliable on-site analysis enables operators to make quick, real-time decisions on further coring, testing and well completion activities.

X-ray Computed Tomography (CT), one of the most widely used imaging techniques, permits visualisation of internal rock features. It was first introduced as a radiological imaging technique by Hounsfield in 1972. A CT scanner is a non-destructive imaging tool that allows a core to be scanned while still contained in the fibreglass barrels and the plastic liner materials. It reveals not only the internal structures of the core but also the damage done by various coring and core-handling actions prior to the laboratory measurements. The results can be used to determine slabbing directions or even optimise the planning of core plug positions [9]. Furthermore, a statistically representative sampling strategy is required for heterogeneous formations where CT scanning can be used to assess the degree of heterogeneity [2]. Figure 2.8 shows a schematic of computer-aided scanning of a core material using X-ray tomography and CT images obtained from a whole-core scanning study.

2.4.2. *Sample Selection*

Core sampling has a significant influence on the success of core-analysis laboratory measurements. A poorly conducted sampling programme may limit the scope and result in ineffective experimental measurements. Sample selection must serve the needs of different disciplines such as geology, petrophysics and reservoir engineering. Ideally, sampling should result in a statistically meaningful representation of the core material. Sample selection varies depending on the type of test. As the requirements of basic core analysis and special core analysis differ significantly, sample selection should be done accordingly to meet the overall core-analysis objectives.

2.4.2.1. *Basic (routine) core analysis sampling*

RCAL is usually done on every foot ($\sim$30 cm). If too many plugs fall in regions of poor quality core material, plugs may be taken at different positions (a few inches away from the predetermined

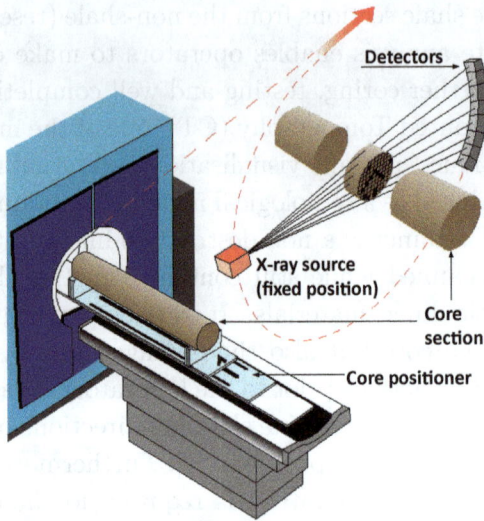

(a)

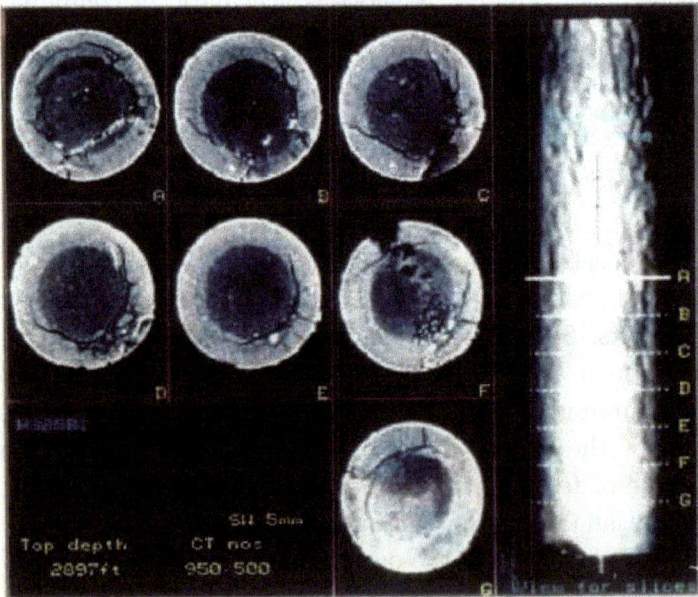

(b)

Figure 2.8. A schematic of computer-aided scanning of a core material (a) [9] and CT images obtained from a whole core scanning study (b) (Courtesy Core Laboratories).

locations). In general, emphasis should be placed on the cutting of plugs as close to the 1 ft spacing as possible, without any regard for variations in lithology. Otherwise, a bias towards apparently better formation properties may be unwittingly introduced, which can lead to improper log calibration [5]. However, plugs that represent two different lithologies (on the border of different lithologies) should be avoided, as the experimental data obtained on these core plugs can be significantly misleading. Core plugs can be cut in orientations both parallel and perpendicular to the bedding. This would help to evaluate anisotropic reservoir parameters such as permeability.

2.4.2.2. *Special core analysis sampling*

As the name suggests, some special precautions should be taken while sampling for SCAL measurements. Unlike RCAL measurements, samples are not taken at regular intervals for SCAL measurements. Most attention is often placed on rock type within the formation. The location and number of samples chosen for SCAL measurements should be representative of rock type under consideration. A core that first appears completely uniform may actually be highly heterogeneous with respect to petrophysical parameters. Therefore, while making the sample selection, use of a non-destructive imaging technique such as CT scanning is highly recommended. It is also recommended that twice as many samples to be taken for a given measurement for the purpose of maximising representativeness of the samples, or repeat measurements may be needed [5]. However, the number of SCAL samples is usually much smaller than that in basic core analysis.

2.4.3. *Plug Preparation*

As mentioned in the previous sections, plug selection is dependent on the type of measurements to be conducted. For basic core analysis, plugging is performed every foot ($\sim$30 cm); however in special core analysis, emphasis is on rock types. Between the time plugs are drilled and core analysis measurements begin, several preparation steps should be taken, such as cleaning, saturation measurement and drying.

2.4.3.1. *Plug drilling*

A whole core is slabbed into two pieces. Prior to the slabbing phase, selected sections of the core may be preserved for whole core analysis or for certain SCAL experiments. The optimum slab plane is usually determined via CT scanning. Thicknesses of the pieces are approximately 1/3 and 2/3 of the entire core along its long axis. Plugs are cut from the thicker section (2/3 piece). While drilling plugs, various lubricants are used depending on the core material. Fresh water is used for clean sands and carbonates. Kerosene (or Blandol) is used for shale and halite-bearing samples. Brine is used for cores containing clays or those from high salinity environments. Unconsolidated cores are often kept frozen and plugs are drilled using liquid nitrogen. Plugging usually takes 10 min–15 min per piece and plug dimensions are 2.5 cm (1″) in diameter and 5 cm–7 cm (2″–3″) in length for most conventional measurements. Plugs for special core analysis measurements are cut usually little larger (1.5″ in diameter). Fluid flow properties may vary with sample orientation. Therefore, care needs to be taken in selecting the direction of the bedding. Horizontal plugs should be drilled parallel to the apparent bedding plane while vertical plugs are drilled perpendicular to the apparent bedding plane [5].

2.4.3.2. *Plug cleaning*

Before porosity and permeability measurements take place, samples should be thoroughly cleaned of reservoir fluids. Cleaning is achieved by a hot solvent extraction (Soxhlet) technique. Depending on the rock type, fluid characteristics, rock mineralogy and timing, the most appropriate solvent is determined. Toluene is the solvent most commonly used to extract water and hydrocarbons. It is usually followed by extraction of the salt with a chloroform/methanol mixture.

The hot solvent extraction (Soxhlet) technique removes all the fluids from the core. Therefore, an initial fluid saturation measurement, if necessary, should be performed before the cleaning process takes place. Dean–Stark distillation is the most widely used

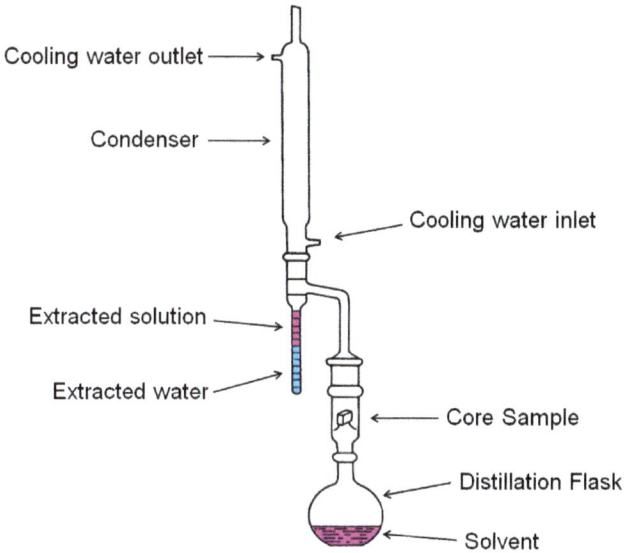

Figure 2.9. Dean–Stark distillation extraction apparatus.

technique for fluid saturation measurements. Figure 2.9 shows a schematic of the Dean–Stark apparatus. The sample is first weighed and placed into the apparatus. Solvent is vaporised by boiling, and it rises up to extract the water from the sample. Solvent and water vapours then condense in a reflux-type condenser and are collected in a calibrated receiving tube. Water is immiscible with the solvent and settles at the bottom of the tube, as it is the denser phase. Therefore, the extracted volume of water can be measured directly in the receiving tube. The solvent refluxes into the distillation flask. The extracted oil remains in solution. Initial water saturation is calculated by dividing the total volume of water collected at the receiving tube into the measured pore volume of the sample. After the oil extraction is completed, initial oil content is determined gravimetrically by using the weight of the sample before and after the distillation process.

Dean–Stark distillation usually takes between 7 and 10 days. It is a non-destructive technique, although some minerals may be affected. Water volume determinations are usually accurate and precise. Equations for the calculation of fluid saturations via the Dean–Stark technique are shown below:

Weight percentage of water

$$= \frac{\text{volume of water} \times \text{density of water} \times 100}{\text{initial weight of sample}} \qquad (2.1)$$

$$\text{Weight percentage of solid} = \frac{\text{dry weight of sample} \times 100}{\text{initial sample weight}} \qquad (2.2)$$

Weight percentage of oil

$$= \frac{\begin{array}{c}\text{initial weight of sample} - \text{dry weight of sample} \\ -\text{weight of water}\end{array}}{\text{initial weight of sample}} \times 100. \qquad (2.3)$$

Note that saturations are expressed in terms of volume percentage of the phase with respect to the total pore volume. Therefore:

$$\text{Water saturation} = \frac{\text{volume of water} \times 100}{\text{pore volume}} \qquad (2.4)$$

$$\text{Oil saturation} = \frac{\text{weight of oil} \times 100}{\text{density of oil} \times \text{pore volume}}. \qquad (2.5)$$

As can be seen from Eqs. (2.4) and (2.5), a saturation calculation requires the pore volume of the sample, which is obtained only after the core is properly cleaned and a measurement is conducted. There may also be a need for volume correction, as the saline water density is not same as the density of the distilled water. Knowing the brine density and salinity, the volume of the brine originally in the sample can easily be calculated.

2.4.3.3. *Drying*

After the saturation measurement is conducted and cleaning is performed, the core is dried. Prior to porosity and permeability measurements, the salt and all the remaining solvent must be removed. There are several drying techniques; oven drying is the most common, most inexpensive and quickest.

Multiple core plugs can be simultaneously dried in a vacuum convection oven. The temperature is set at around 95°C, and each core sample should be dried until constant weight is obtained. When hydrated minerals such as clays are present, humidity ovens may be used to minimise sample alteration. Humidity ovens can be set at 60°C and 40% relative humidity. Because of the low temperature, drying may take several days. The resulting "effective" porosity needs careful calibration to relate with either log-derived effective or total porosity [5].

If a rock contains minerals such as hairy illite, which are sensitive to fluid phase changes, then the Critical Point Drying (CPD) technique may be employed. Certain minerals can be damaged due to the large interfacial forces created in tiny cavities. The CPD technique prevents the development and advancement of a gas–liquid or liquid–liquid interface within the rock by raising the fluid above its critical point. However, this cannot be easily achieved with the original fluids (oil and brine) in the core. Therefore, oil and brine are replaced from the pore space by methanol and liquid CO_2 through diffusion; CO_2 is preferred because of its critical properties (32°C and 72 bar). Methanol is used as an intermediate liquid to ensure full miscibility [5]. In order to investigate the drying effects, Scanning Electron Microscopy (SEM) is generally used in conjunction with CPD. This technique is usually slow as it depends on the diffusion time, which is a function of sample size, permeability and fluids initially in place. Critical point drying can take from two weeks to two months.

2.5. Basic Core Analysis Laboratory Measurements

Basic core analysis is sometimes referred to as routine core analysis. However, in the last few years, the word *routine* has intentionally been replaced by *basic* in order to emphasise that nothing related to core analysis is routine. Instead, for each new reservoir every step requires extra care and special attention, from coring all the way to data reporting. As the name suggests, basic core analysis includes the measurements of basic physical properties. These are grain density, porosity, permeability and fluid saturation.

2.5.1. *Porosity and Grain Density Measurement*

Porosity, a measure of space available for hydrocarbon storage, is one of the most important parameters for the development of petroleum reservoirs. It is defined as the void (pore) volume of the sample divided by its bulk volume.

There are several different methods developed for porosity measurements. These methods calculate three critical parameters: (1) bulk volume, (2) grain volume and (3) pore volume.

2.5.1.1. *Bulk volume*

Although the bulk volume may be computed from measurements of the dimensions of a regularly shaped sample, the usual procedure utilises the observation of the volume of a fluid displaced by the core [4]. This technique is particularly advantageous as it can also be applied to irregular-shaped samples. However, care should be taken to prevent fluid invasion into the pore space of the rock. This is usually achieved by using mercury as the fluid to be displaced. Due to its non-wetting characteristics, mercury tends to stay away from the pore space of the rock. Special care must be taken while applying this technique on samples, since the sample may be contaminated by mercury invasion into large void areas.

2.5.1.2. *Grain volume*

The most widely used technique is the Boyle's law (gas expansion) method. This method involves the expansion of a compressed gas (usually helium, due to its small molecular size and low adsorption on rock surfaces) into a clean, dry sample. A schematic of a Boyle's law porosimeter is shown in Figure 2.10. Gas (helium) is initially admitted to the reference cell of known volume at a pre-set reference pressure (V_2 and P_2 in Figure 2.11). The gas in the cell then expands into a connected chamber of known volume and pressure (V_1 and

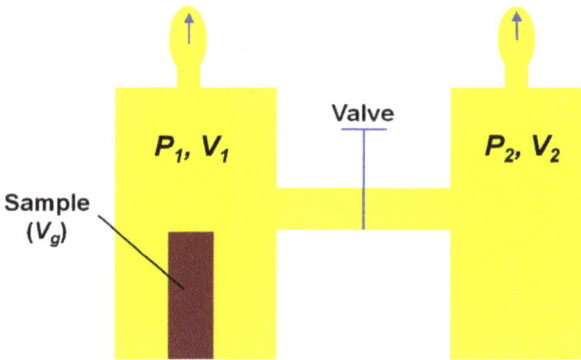

Figure 2.10. A schematic of the double-cell Boyle's law porosimeter.

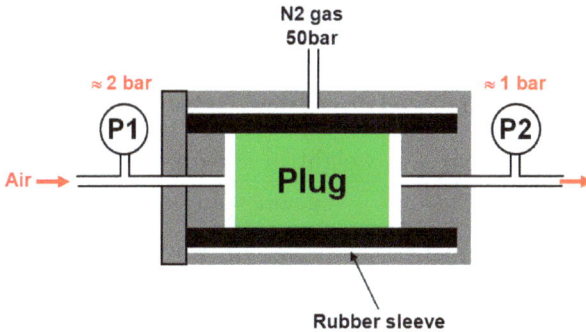

Figure 2.11. A schematic of a Ruska permeameter.

P_1), which contains the core sample of unknown grain volume. The equilibrium pressure is then measured and grain volume (V_g) can be calculated from the equation

$$P_1(V_1 - V_g) + P_2 \cdot V_2 = P(V_1 + V_2 - V_g), \qquad (2.6)$$

where P_1 and P_2 are the measured pressure values at cell 1 and 2 before the valve is open and P is the equilibrium pressure of the whole system after the valve is opened.

2.5.1.3. *Pore volume*

Although it can be measured directly, pore volume is often calculated indirectly by subtracting the grain volume from the bulk volume.

All the direct measurement techniques yield effective porosity. The methods are based on either extraction of a fluid from the rock or introduction of a fluid into the pore space of the rock [4].

2.5.1.4. *Grain density*

A pycnometer, which is a small glass flask of known weight, accurately calibrated for volume, can be used to measure grain (matrix) density. A dry and clean sample is placed in a pycnometer. After weighing, the pycnometer is filled with toluene or kerosene, and the solvent is degassed. The weight of pycnometer, sample, and solvent is then determined at a known temperature. Grain density can then be calculated from the weights, pycnometer volume and solvent density [5].

Grain density can also be calculated by simply weighing a dry sample on a precise balance and dividing the measured weight into the grain volume calculated from Boyle's law (if available):

$$\text{Grain density} = \frac{\text{weight of the dry sample (g)}}{\text{grain volume (cm}^3)}. \tag{2.7}$$

Laboratory measurement techniques of porosity have not changed significantly in the last few decades. The accuracy of the measurement may be affected by several factors, such as grain volume measurement accuracy and grain losses in some cases. When comparing experimentally measured porosities with log-derived ones, effects of sample preparation and representativeness of sample volume should be taken into account.

2.5.2. *Permeability Measurement*

Permeability is defined as the ability of a formation to transmit fluids. Darcy's law states that an incompressible single-phase fluid flows through the pore space of the rock with a volumetric flow rate of

$$Q = \frac{kA(P_i - P_o)}{\mu L}, \tag{2.8}$$

where Q is the volumetric flow rate (cm^3/s) under laminar flow, k is the permeability (Darcy), A is the cross-sectional area (cm^2),

P_i and P_o are inlet and outlet pressures (atm), μ is viscosity (cp) and L is length (cm). Note that the unit of permeability is termed as the Darcy. The Darcy is used in oil field units and 1 Darcy is approximately $10^{-12}\,\text{m}^2$.

Accurately predicting permeability and accounting for its heterogeneity are crucial in field development studies. The most direct method of permeability estimation is conducting experiments using cores ranging from sidewall plugs to full diameter cores. Well-test analysis also gives information about fluid mobility (hence permeability) at a larger volume scale, if the test is designed properly and the interpretation model is well constrained by geological and geophysical data.

2.5.2.1. *Steady-state measurement*

A geometrically regular sample of known length and diameter is placed into a Hassler-type core holder. Gas, which is typically air or nitrogen, is injected from the inlet and flows through the sample by creating a pressure gradient between the inlet and the outlet. Pressure gradient across the core plug and the flow rate are measured with a permeameter. A schematic of a Ruska permeameter is illustrated in Figure 2.11. The permeability is calculated using a modified form of Darcy's equation that takes gas compressibility into account:

$$K_{\text{gas}} = \frac{2{,}000 \cdot P_a \cdot \mu \cdot Q \cdot L}{(P_1^2 - P_2^2) \cdot A}, \tag{2.9}$$

where K_{gas} is permeability to gas (mD), P_a is atmospheric pressure (atm), μ is viscosity (cp), Q is the volumetric flow rate (cm^3/s), L is the length (cm), P_1 and P_2 are inlet and outlet pressures (atm) and A is the cross-sectional area (cm^2). Note that 1 millidarcy (mD) is equal to $1/1{,}000$ of a Darcy.

For detailed analysis of spatial permeability variation in the core, a probe permeameter (mini-permeameter) can be used. A schematic of a probe permeameter is shown in Figure 2.12. In the probe permeameter, gas flows from the end of a small-diameter probe that is sealed against the surface of a slabbed or exposed core.

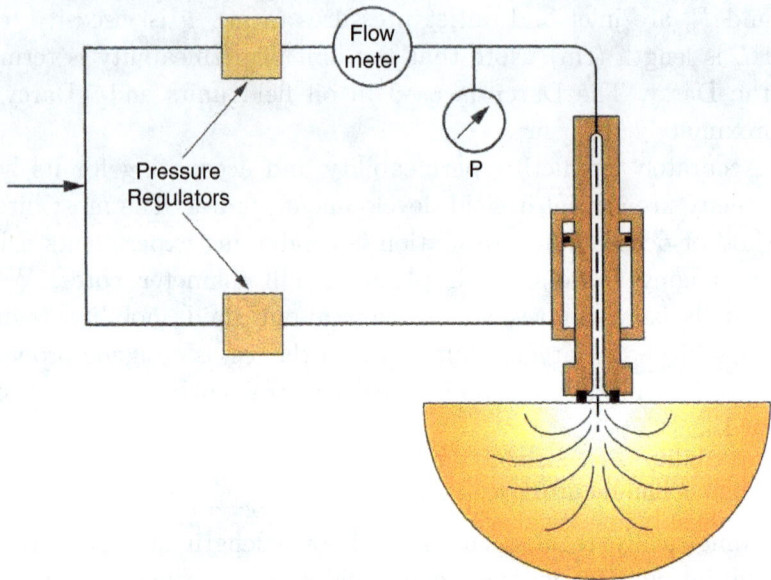

Figure 2.12. A schematic of a probe permeameter (mini-permeameter) [5].

The gas flowrate and the pressure in the probe are measured and used for permeability calculation. As the measured permeability is localised to a small region near the seal, this technique is particularly useful for characterising small-scale geological heterogeneities. Typically, the probe permeability measurement results in a permeability profile along the core or a 2D array across a slabbed core face. Obtained data can be calibrated against the core plug permeability measurements.

2.5.2.2.   *Unsteady-state measurement*

With the increase in the computational power and accuracy of pressure transducers, unsteady-state permeability measurements are becoming more common. Figure 2.13 shows an illustration of the pulse-decay permeameter. A dry sample is placed into the core holder. A pressure pulse is introduced by increasing the pressure in the upstream vessel. The system is then allowed to return to equilibrium pressure, and the rate of approaching equilibrium

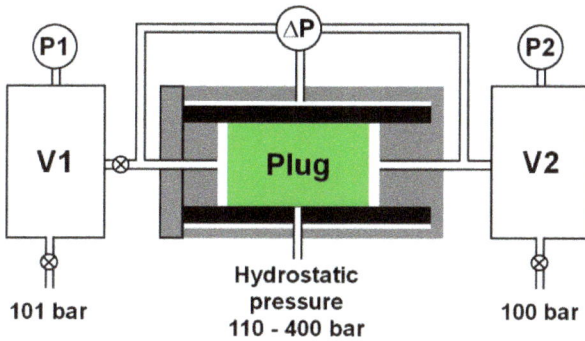

Figure 2.13. A schematic of a pulse-decay permeameter.

depends on the permeability of the core sample. However, in the pulse-decay method, it is not necessary to wait until the pressure equilibrium is reached. Therefore, this technique (or one of its variations) is particularly useful for low permeability samples (below 0.1 mD), where steady-state measurement may face the problem of prohibitively lengthy time to reach pressure/flow equilibrium.

2.5.2.3.   *Whole-core measurement*

The whole core is prepared in the same manner as the plugs. It is then placed into a rubber sleeve of a core holder where it is subjected to a confining pressure of around 20 bar. This provides sealing along the sides of the sample. A schematic of a whole-core measurement apparatus is shown in Figure 2.14. The principle is the same as in the steady-state measurements; vertical permeability can easily be determined by flowing gas through the length of the core. Horizontal permeability measurement, however, is more complex. Gas flows through the cylindrical surface of the sample using an array of permeable screens, which cover opposite quadrants on the surface and are rotated through 90° so that the measurement can be carried out in two perpendicular orientations. The higher of the two horizontal permeabilities is referred to as k_{max} and the lower as k_{90}. Typically, a geometric mean permeability is calculated, which is used in porosity-permeability correlations and in log calibration for permeability estimation [5].

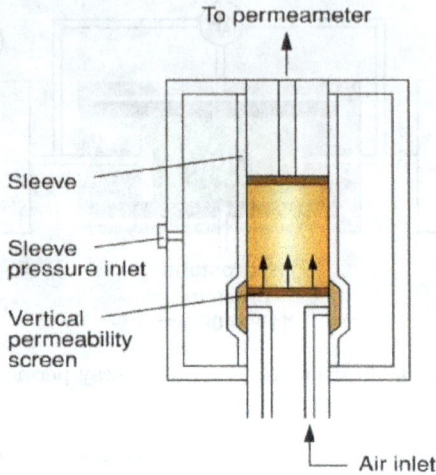

Figure 2.14. A schematic of whole-core permeability measurement apparatus [5].

2.5.2.4. *Factors affecting permeability measurement*

Permeability of a core sample can be affected by a variety of phenomena. Gas slippage, confining pressure, pore content, inertial (turbulence) effects and improper plug cleaning/drying are a few examples. Therefore, certain precautions should be taken while preparing the sample and conducting the measurement.

2.5.2.4.1. Gas slippage

Klinkenberg [10] has reported variations in measured permeability when comparing data, which were collected using non-reactive liquids versus gases. These variations were associated with a laboratory effect called gas slippage, which can be described as the ability of a gas molecule to more easily retain forward velocity along a solid interface compared to a liquid molecule.

Liquid velocities normally approach zero at the solid wall. However, gas molecules have a non-zero wall velocity. This may result in two different measurements of rock permeability depending on the fluid used in the experiment. Since permeability is a rock property, it should be independent of the fluid injected into the system. Therefore, data obtained through gas injection should be corrected.

Klinkenberg showed that measured permeability values are higher at low mean (average) pressures, since gas molecules do not adhere to the pore walls as liquid molecules do and hence the slippage of gases along the pore walls occurs. This so-called slippage effect decreases with increasing pressure, as a gas begins to act like a liquid once the flowing pressure increases. Experimental data show that a plot of the reciprocal of mean flowing pressure with respect to permeability yields a straight line. The straight line can be extrapolated to infinite mean pressure and the permeability value at the extrapolated point is referred as the equivalent liquid permeability. A graphical illustration is shown in Figure 2.15. The mathematical expression of this phenomenon is called the Klinkenberg equation and can be written as

$$K_{\mathrm{l}} = \frac{K_{\mathrm{g}}}{\left(1 + \frac{b}{P_{\mathrm{m}}}\right)}, \tag{2.10}$$

where K_{l} is the equivalent liquid permeability, K_{g} is gas permeability, P_{m} is the mean flowing gas pressure and b is Klinkenberg constant, which is dependent on the type of the gas and rock.

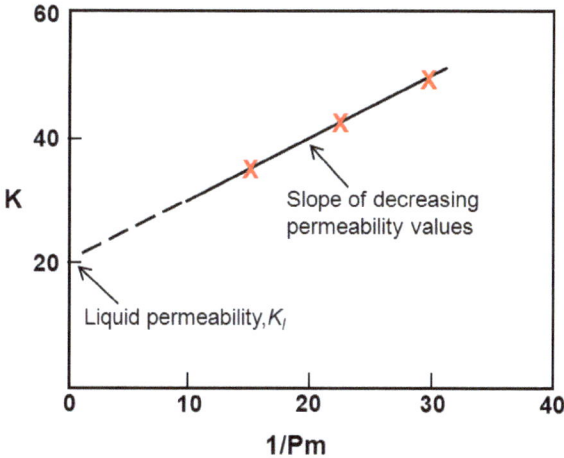

Figure 2.15. An illustration of the Klinkenberg permeability correction. Note that the crosses are the measured permeability values at different pressures. The straight line is extrapolated to infinite mean pressure (reciprocal converges to zero) and the permeability value is recorded as the equivalent liquid permeability.

Note that the correction factor, on a percentage basis, is greater for low-permeability and low-pressure conditions. It becomes smaller as the permeability value and the pressure increase. It is usually negligible at reservoir conditions due to large fluid pressure.

2.5.2.4.2. Confining pressure

Ideally, permeability measurements should be conducted by applying a net overburden pressure in order to simulate the reservoir conditions. It has been found that overburden pressure reduces the measured permeability values. The impact is observed more on unconsolidated rocks and on samples containing fractures and microcracks. In order to better understand and account for its impact, a series of confining stress measurements should be performed on selected samples.

2.5.2.4.3. Inertial effects

If the gas flow rate through the core is significantly high, inertial effects may become important and should be accounted for. They usually generate additional pressure drop and may result in underestimation of permeability if not properly accounted for. Although it is often evident in laboratory studies, turbulence effects are in general only applicable in the near-wellbore region of high-rate gas or light-oil reservoirs. High-production gas wells have demonstrated the need for extended core analysis that provides a non-Darcy flow coefficient [11]. Therefore, the Forchheimer correction factor, β, is commonly reported along with the gas permeability measurements in these kinds of specific circumstances. The general form of the Forchheimer [12] equation is

$$dP/dx = \alpha \cdot \mu \cdot v + \beta \cdot \rho \cdot v^2, \qquad (2.11)$$

where P is pressure, x is length, α is the averaging coefficient, which is equal to the inverse of permeability, ρ is density of the gas and v is velocity.

2.5.2.4.4.   Reactive liquids

Although water is considered to be a non-reactive fluid, its interaction with clay minerals may make it behave in a reactive manner in terms of its impact on permeability determination. Fresh water may cause significant clay swelling [13]. Therefore, extra care should be taken while choosing the proper core plugging, cleaning and drying techniques. For example, if the rock contains significant amount of clay minerals, fresh water should be avoided while drilling a plug from the core; instead, brine with an appropriate salinity level and ion balance should be used.

Permeability reduction may also be encountered due to particle movement, which is a function of flow velocity and fluid density. Critical velocity can be used to determine the maximum displacement rate for experimental studies in order to avoid the initiation of the movement of fine particles [2].

2.5.2.5.   *Average permeability of a combination of layers*

Although we assign one single value to each grid block in reservoir simulation studies, the rock is seldom very uniform. Grid block sizes are usually taken in the order of tens to hundreds of meters in the horizontal direction and a few to tens of meters in the vertical direction. Most reservoir rocks will have permeability variations on a much smaller scale than the above-mentioned distances. The reason we use more frequent grid blocks in the vertical direction is that most reservoirs are layered, with each layer having a different permeability. Depending on the direction of flow (either horizontal or vertical), it may be possible to define an effective permeability by one of several averaging techniques.

2.5.2.5.1.   Horizontal flow (parallel to the layering)

If the flow occurs in the horizontal direction (see Figure 2.16(a)) parallel to the layering scheme, then the effective permeability for flow is the weighted arithmetic mean of the individual permeabilities,

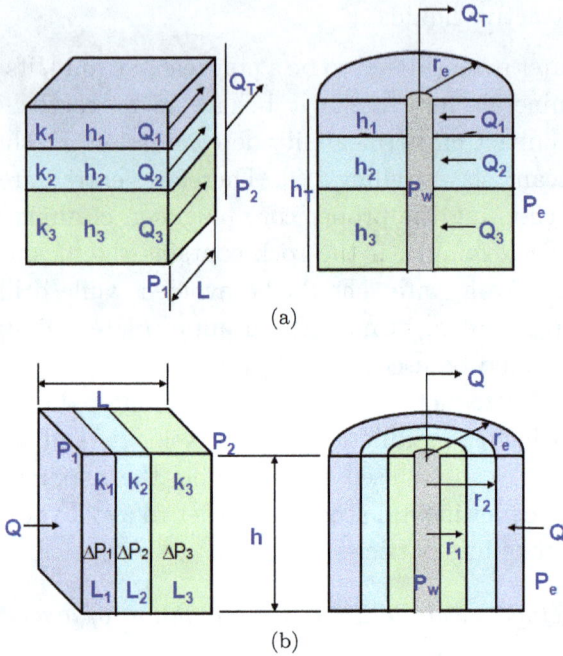

Figure 2.16. Linear and radial flow in (a) parallel combination and (b) series combination of beds.

weighted by the thickness of each individual layer:

$$k_{\text{eff}} = \sum_{j=1}^{n} k_j h_j \bigg/ \sum_{j=1}^{n} h_j, \qquad (2.12)$$

where k_{eff} is the effective permeability, k_j is the permeability for each individual layer, n is the number of layers and h_j is the thickness of each layer.

2.5.2.5.2. Vertical flow (perpendicular to the layering)

If the flow occurs in the vertical direction (see Figure 2.16(b)) perpendicular to the layering scheme, then the effective permeability is the weighted harmonic mean of the individual permeabilities:

$$k_{\text{eff}} = L \bigg/ \sum_{j=1}^{n} L_j/k_j, \qquad (2.13)$$

where L is the total length of the each individual layer L_j. As can be calculated from Eqs. (2.12) and (2.13), the effective permeability for flow parallel to layering is controlled by the most permeable layer, whereas for flow perpendicular to layering, it is controlled by the least permeable layer.

2.6. Special Core Analysis Laboratory Measurements

A special core analysis programme typically involves the following experimental measurements: electrical properties (such as formation factor and resistivity index), wettability, capillary pressure, relative permeability and porosity/permeability measurements at elevated temperatures and pressures. We discuss each measurement technique in detail below.

2.6.1. *Electrical Properties*

Core materials, with the exception of certain clay minerals, are non-conductors. Therefore, electrical properties are mainly determined by the fluids that fill the pore space of the rock. Petroleum reservoirs typically contain oil, water and gas — oil and gas being the non-conductors and water being the conductor. Water residing in porous reservoir rocks contains dissolved salts that allow conduction of electrical currents. The resistivity of a material is the reciprocal of conductivity.

Archie [14] showed that the resistivity of the rock, which is saturated with brine, increases linearly with the resistivity of brine. He defined the proportionality constant as formation factor, which can be estimated as

$$F = \frac{R_o}{R_w} = \emptyset^{-m},\qquad(2.14)$$

where F is the formation factor, R_o is the resistivity of brine saturated rock, R_w is the brine resistivity, ϕ is the porosity and m is the cementation exponent which is obtained using a bi-logarithmic plot of formation factor versus porosity (see Figure 2.17).

Archie derived the above-mentioned relationship between formation factor and porosity using the sandstone cores from Gulf Coast

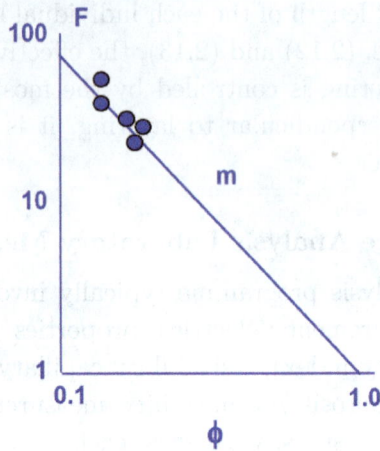

Figure 2.17. An illustration of the estimation of the cementation exponent (m). The cementation exponent is the negative slope of a log-log plot of formation factor (F) versus porosity (ϕ), which generally lies between 1.1 and 2.4 [2]. Note the dots; each represents a separate core sample.

reservoirs. The value obtained for the cementation exponent was around 2. A more generalised form of the Archie equation can be used where formation factor is a function of internal geometry of the rock as well as the porosity. It can be shown as

$$F = \frac{a}{\phi^m}, \tag{2.15}$$

where a is a function of tortuosity of the rock and lies between 0.62 and 3.7 [2].

If the core is partially saturated with brine and hydrocarbon, the resistivity of the rock increases, as brine is the only electrical conductor. Using the experimental data reported in the literature, Archie proposed the following relationship:

$$I = \frac{R_t}{R_o} = \frac{R_t}{F R_w} = \frac{1}{S_w^{-n}}, \tag{2.16}$$

where I is the resistivity index, R_t is the resistivity of rock which is partially saturated with brine, R_o is the resistivity of rock which is fully saturated with brine, F is the formation factor, R_w is the

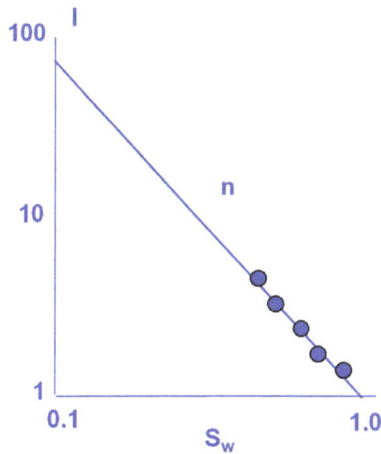

Figure 2.18. An illustration of the estimation of the saturation exponent (n). The saturation exponent is the negative slope of a log-log plot of resistivity index (I) versus water saturation (S_w). Note the dots; each point is measured on the same core sample at different water saturation.

brine resistivity, S_w is the water saturation and n is the saturation exponent, which can be obtained using a bi-logarithmic plot of resistivity index versus water saturation (see Figure 2.18).

In his earlier work, Archie correlated experimental data with the above-mentioned relationship. He found the saturation exponent to be constant and equal to 2 for the tested samples. However, in later studies, a wide range of saturation exponent values has been reported [2].

The purpose of resistivity index measurements is to provide a relationship between resistivity index and water saturation, which can then be used for the interpretation of resistivity logs. As can be seen, the Archie equation (Eq. (2.16)) assumes a linear relationship between resistivity index and water saturation on a log-log scale. However, experimental data sometimes show a non-linear relationship for complex systems. The reason for a non-linear log-log relationship between resistivity index and water saturation can be excessive clay content of the rock, wettability of the system or multimodal pore size distribution. Figure 2.19 illustrates typical resistivity index versus saturation relationships on a log-log plot. The

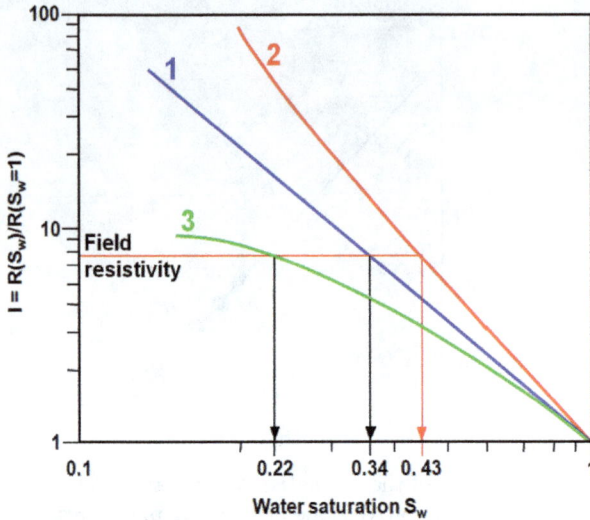

Figure 2.19. Typical I–S_w relationships. A linear relationship is obtained for clean rocks (1). However, an upward curvature may be observed for oil-wet cores (2) and a downward curvature is expected for shaly sands (3).

Archie relationship is only valid for clean rocks without conductive solid minerals (clays), whereas a downward curving relationship is expected for sandstones with high clay content or for shaly sandstones.

For shaly sandstones, especially under low brine salinities, the Archie equation may no longer be valid. This is because of the existence of electrical double layers in shaly sands, which provide a second conductive path for electrical current [2]. Therefore, a downward curving log-log relationship between the resistivity index and water saturation is typical for most shaly sandstones (see curve number 3 in Figure 2.20). A precise knowledge of this curvature is necessary for proper interpretation of the data obtained from wire-line logs. A number of shaly sand models have been developed and are being used. Waxman–Smits [15] is one of the most commonly used techniques and can be represented as

$$R_t = \frac{R_w \cdot \phi^{-m^*} \cdot S_w^{-n^*}}{(1 + R_w \cdot B \cdot Q_v \cdot S_w)}, \tag{2.17}$$

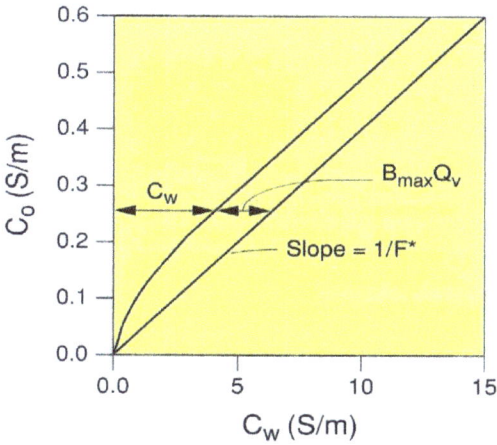

Figure 2.20. An illustration of brine saturated rock conductivity (C_o) versus brine conductivity (C_w) plot for a shaly sandstone [5].

where m^* and n^* are intrinsic Archie exponents, B is the equivalent conductance of clay-exchange cations and Q_v is the cation exchange capacity per unit volume.

A number of methods are available for measuring cation exchange capacity per unit volume, Q_v. The conductometric titration method, membrane potential and multiple salinity measurements are a few of the examples. In the conductometric titration method, a cleaned sample is crushed and weighed. The exchangeable cations are exchanged using a suitable leaching solution and are quantified by conductometric titration; this involves titrating the solution while monitoring the conductivity, which displays a rapid increase after the correct titration has been reached [5]. The conductometric titration technique is also known as the wet-chemistry method. It may not be recommended for an accurate Q_v measurement, as the test is destructive. It involves crushing the sample, hence destroying the clay morphology and distribution and creating additional ion-exchange sites [2].

In a membrane potential measurement, an electrical potential is generated when two brines of different salinity are brought into contact. When a shaly sample is positioned at the interface, the electrical potential increases. The magnitude of the increase is directly related to the cation exchange capacity per unit volume, Q_v [5]. This technique is applicable to both consolidated and unconsolidated rocks. It is a non-destructive measurement, which properly incorporates the effects of clay distribution within the sample.

In a multiple salinity measurement, the conductivity of a brine-saturated rock sample, C_o, is measured at different brine salinities of known brine conductivity, C_w. Four brines are usually used, starting with the lowest salinity. The sample is flooded with brine until the conductivity reaches equilibrium for different salinities. The quantity BQ_v, which is called the clay conductivity in the Waxman–Smits equation (Eq. (2.17)), is determined from the plot of C_o versus C_w (see Figure 2.20). The clay corrected formation resistivity factor, F^*, is determined by fitting a straight line of slope $1/F^*$ through the high salinity data. Intrinsic Archie exponents m^* and n^* can be determined through correlations of F^* with porosity. Clay conductivity, BQ_v, is obtained from the difference between F^*C_o and C_w.

The equivalent conductance of clay-exchange cations, B, which is used to scale the cation exchange capacity per unit volume, Q_v, to units of conductivity, can be calculated as a function of brine resistivity, R_w, and temperature, $T(°C)$. Although several different correlations have been suggested, the one developed by Juhasz [16] is the most widely used relationship:

$$B = \frac{-1.28 + 0.225T - 0.0004059T^2}{1.0 + R_w^{1.23}(0.045T - 0.27)}. \tag{2.18}$$

The most common technique for determining resistivity index is the porous plate method. A clean and dry plug is placed in a Hassler-type pressure cell and saturated with brine under low confining stress. The sample is saturated with 100% brine, and the resistivity, R_o, is measured. If the system is strongly water-wet, air can be injected as the invading phase, which represents only the drainage process. Otherwise, oil/brine displacement may become necessary. For a given injection pressure, sample resistivity and volume of expelled brine are monitored when equilibrium is reached (meaning that no more

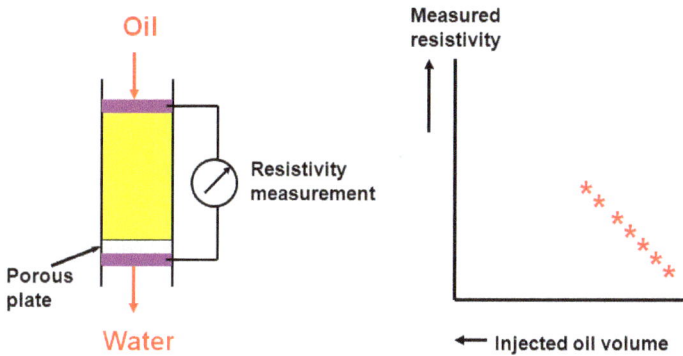

Figure 2.21. An illustration of the porous plate method for resistivity index measurement. Note that porous plate is located below the core.

brine is produced). The measurement is repeated for increasing steps in oil pressure. In order to mimic the imbibition process, brine may be injected to displace oil. An illustration of the porous plate measurement is shown in Figure 2.21, where the saturation is calculated from volumetric measurements at each capillary pressure level when equilibrium is reached. The porous plate technique can also be performed under a representative confining pressure. Since the sample is kept in place under pressure throughout the test, there is no risk of grain losses and the capillary contact between the core and the plate is maintained. A by-product from such an experiment is the capillary pressure curve.

Another commonly used method for determining resistivity index is the continuous injection technique. A clean and dry sample is first weighed. It is then 100% saturated with brine, and the resistivity, R_o, is measured. Oil phase is injected into the sample at a constant slow injection rate. The injection rate is chosen so that a pore volume of oil would be injected in about two weeks. Water saturation is determined continuously from injected/produced fluid volumes through a material balance calculation. The advantage of the continuous injection technique over the porous plate method is that the measurement is faster, since reaching pressure equilibrium is not required at each saturation point. However, the disadvantage of this technique compared to the porous plate method is that the capillary

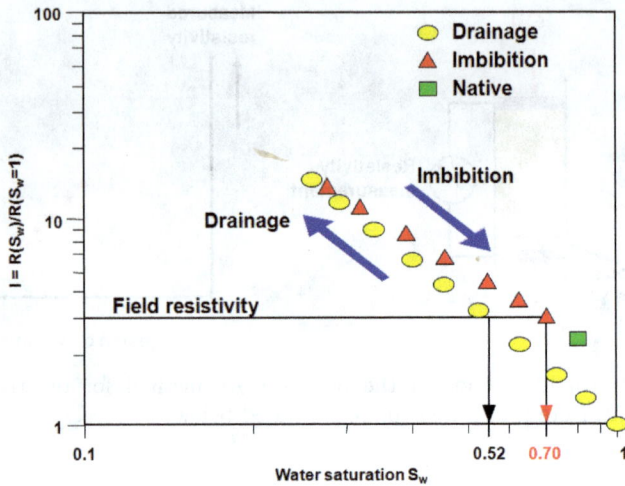

Figure 2.22. Effect of hysteresis in I–S_{w} relationship.

pressure curve cannot be obtained, as the pressure equilibrium is not reached at the measured saturation points.

It is important to take hysteresis into account while performing resistivity index measurements. Usually, measurements are conducted while the non-wetting phase is displacing the wetting phase (during drainage, i.e. oil displacing water cycle). However, the same data points may not be obtained during the imbibition cycle (i.e. water displacing oil), since the wettability of the system may alter after oil invasion (see Figure 2.22). Therefore, before conducting a measurement on the imbibition cycle, special care should be taken to reach the proper wetting state, which will be discussed in more detail in the next section.

2.6.2. *Wettability*

Petroleum reservoirs consist of water, oil and gas. When dealing with multi-phase systems, it is necessary to take the wetting characteristics of different fluids into account. Wettability, sometimes called the energy of adhesion, is defined as the ability of a fluid to spread on a solid surface in the presence of other immiscible fluids [17].

For an oil–brine system, wettability is the preference of the rock for either oil or water. Basic core analysis experiments or the experiments considering only the behaviour in primary drainage (non-wetting phase injection) can be conducted on cleaned core samples. However, when considering the behaviour on the imbibition cycle (wetting phase saturation increasing), wettability becomes a determining factor of flow behaviour. Wettability controls the location, flow behaviour and distribution of the fluids in oil reservoirs. The wettability of a core will affect almost all types of special core analysis, including capillary pressure, relative permeability and electrical properties as well as the waterflood behaviour and tertiary recovery. The most accurate results are obtained when experiments are performed with native-state or wettability-restored cores and representative crude oil and brine at reservoir temperature and pressure. Such conditions provide cores that mimic wettability as found in the reservoir [18].

When two fluids are in contact with a solid, the angle between the solid and the interface of two fluids is called contact angle, θ. Wettability is often quantified by this contact angle. Contact angle, by convention, is measured through the denser phase. If it is less than $90°$, then the system is considered to be wetting to the denser phase. Otherwise, if the contact angle θ is greater than $90°$, then it is non-wetting to the denser phase. Figure 2.23 illustrates a water-wet system, where the contact angle, which is measured through water as being the denser phase, is less than $90°$.

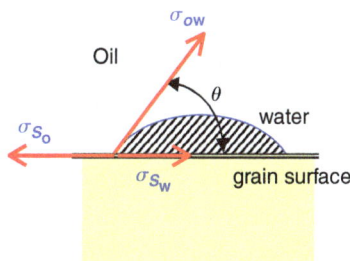

Figure 2.23. An illustration of a water-wet rock surface. Note that the contact angle is less than $90°$ in this case.

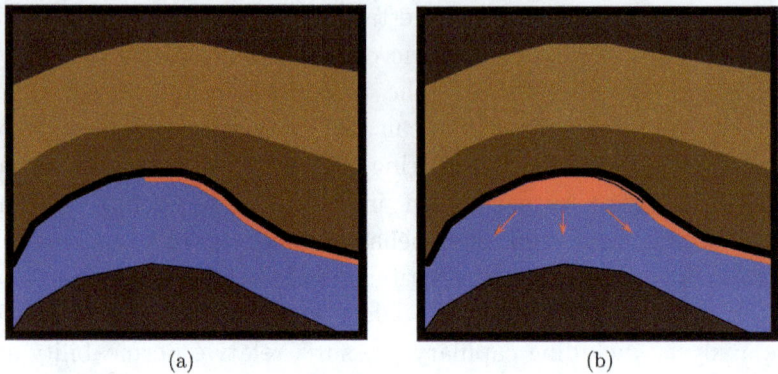

Figure 2.24. The reservoir is initially strongly water-wet (a). However, once oil migrates into the reservoir, oil displaces water, ageing takes place and the reservoir becomes more oil-wet or mixed-wet (b).

Oil reservoirs are initially believed to be strongly water-wet because almost all clean sedimentary rocks are strongly water-wet. They are initially saturated with water, and oil migration takes place at a later stage. However, once oil migrates into the system, water-wet behaviour of the rock may change by adsorption of polar compounds and deposition of organic matter originally in the crude oil [17, 18]. Figure 2.24 illustrates the change in the original wettability of the system. Once oil migration takes place, the system's wettability may remain water-wet or can become intermediate-wet, mixed-wet or oil-wet. Intermediate (or neutral) wettability means that rock has no preference for either oil or water. In other words, the contact angle is around 90°. Mixed (or fractional) wettability means that certain parts of the reservoir are water-wet; the rest are oil-wet.

Alterations in wettability may have a significant impact on the results of most core analyses. Ideally, carefully preserved native-state cores are preferred for multi-phase flow experiments where alterations to the wettability of the reservoir rock are minimised. If single-phase experiments are conducted, such as porosity/permeability measurement where wettability of the rock is not critical, then cleaned cores can be utilised. While cleaning the cores, extra effort is made to remove all the fluids and adsorbed organic material by injecting solvents through the cores. The third type is a restored-state core in which the core is first cleaned, and then saturated with brine, followed by oil injection. The core is then aged at reservoir temperature for 4 to 6 weeks to restore its native wettability state [18].

When dealing with multi-phase systems, the effects of the forces acting at the interface of the immiscible fluids should be considered. In an oil–brine system, the force acting at the oil and brine interface is called water–oil interfacial tension, σ_{ow} (see Figure 2.23) and can be related to the contact angle (θ) as

$$\sigma_{ow} = \frac{\sigma_{so} - \sigma_{sw}}{\cos \theta}, \tag{2.19}$$

where σ_{so} is the surface tension between the solid and oil, σ_{sw} is the surface tension between the solid and water, and θ is the contact angle which is measured through the denser phase.

There are a number of techniques used to measure wettability in the laboratory. These methods include direct measurement of contact angle or the utilisation of capillary pressure data. The two most widely used wettability measurement techniques are the Amott–Harvey index and the United States Bureau of Mines (USBM) method. These methods involve the measurement of spontaneous and forced imbibition capillary pressure curves which will be discussed in more detail in the next section. Figure 2.25 illustrates the measurement of wettability indices using Amott–Harvey and USBM methods.

The Amott–Harvey index is measured as the ratio of the volume of fluid displaced by spontaneous imbibition to the volume of fluid displaced by both spontaneous and forced imbibition. The Amott water wetting index, I_w, is calculated as the ratio of the amount of water spontaneously imbibed to the total amount of water imbibition, including the spontaneous and forced (volume shown as line $p - q$ divided by the volume shown as line $p - s$ in Figure 2.25). The Amott oil wetting index, I_o, is calculated as the ratio of the amount of oil spontaneously imbibed to the total amount of oil imbibed (volume shown as line $s - r$ divided by the volume shown as line $s - p$ in Figure 2.25). The Amott–Harvey index is then calculated by

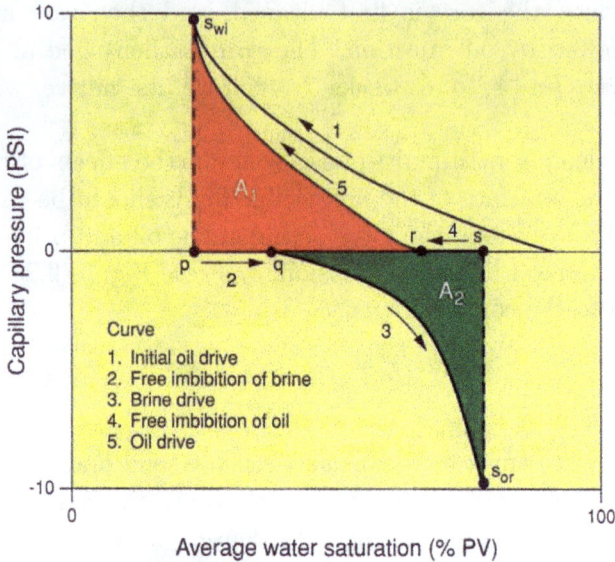

Figure 2.25. Amott–Harvey and USBM wettability indices [5].

subtracting the Amott oil-wetting index, I_o from the Amott water-wetting index, I_w:

$$AI = I_w - I_o, \tag{2.20}$$

where AI stands for the Amott–Harvey index, which is larger than zero for water-wet, close to zero (between -0.3 and 0.3) for intermediate-wet, and smaller than zero for oil-wet systems. Note that the larger the absolute value of AI, the greater the wetting preference will be.

Wettability measurement with the USBM technique is similar in principle to the Amott method. However, it relies on the areas under the capillary pressure curves. The USBM wettability index, W, is calculated as the logarithm of the ratio of the area under the oil injection capillary pressure (A_1 in Figure 2.25) to the area under the water injection capillary pressure (A_2 in Figure 2.25):

$$W = \log \left(\frac{A_1}{A_2} \right), \tag{2.21}$$

where W stands for the USBM wettability index and is measured as a positive value for water-wet, about zero for intermediate-wet and a negative value for oil-wet systems. The larger the absolute value is, the greater the wetting preference.

2.6.3. *Capillary Pressure*

Capillarity has two important effects in petroleum reservoirs: it is responsible for the initial distribution of the fluids in a reservoir under capillary–gravitational force balance and it is the control mechanism whereby oil and gas move through reservoir pore spaces until they are confronted with a barrier [19].

Capillary pressure, P_c, is defined as the differential pressure between two immiscible fluids. In an oil–brine system, capillary pressure is generally defined as the pressure difference between the oil phase and the water phase:

$$P_c = P_o - P_w, \qquad (2.22)$$

where P_o is the pressure of the oil phase and P_w is the pressure of the water phase. Note that the pressure difference, P_c, is generally expressed as a function of water saturation, S_w.

Capillary pressure depends on both rock and fluid properties. Interfacial tension, wettability and pore-size distribution are the key parameters to determine capillary pressure. Capillary pressure between two immiscible fluids in a circular cross-sectional pore element is represented by the Young–Laplace equation as

$$P_c = \frac{2\sigma \cos \theta}{r}, \qquad (2.23)$$

where σ is the interfacial tension between the fluids, θ is the contact angle and r is the mean pore radius.

It is worthwhile to mention that capillary pressure is strongly dependent on the saturation history of the system. From a reservoir engineering perspective, both drainage and imbibition capillary pressure curves may be required for integrated reservoir studies, depending on the history of saturation change in the reservoir. From a petrophysicist perspective, P_c is an important parameter, which can be used to evaluate the field-wide variation of water saturation with respect to the height of the reservoir.

2.6.3.1. *Capillary rise*

If we consider a case where a clean and water-wet capillary tube with a circular cross-section and a small diameter is placed in a large open vessel containing oil and water, then the water level would rise in the capillary tube above the height of the water level in the large vessel. This rise in the height is due to the capillary forces. The water level would rise in the tube until hydrostatic equilibrium is reached (see Figure 2.26). The capillary force acting to pull the water upward is balanced by the force due to weight of the water column in the tube. By equating the two quantities, we reach the relationship

$$(\rho_{\mathrm{w}} - \rho_{\mathrm{o}})gh = \frac{2\sigma \cos \theta}{r}, \tag{2.24}$$

where ρ is the fluid density, g is the acceleration due to gravity, h is the column height, σ is the interfacial tension between oil and water, θ is the contact angle and r is the radius of the capillary tube. The subscripts w and o stand for water and oil, respectively.

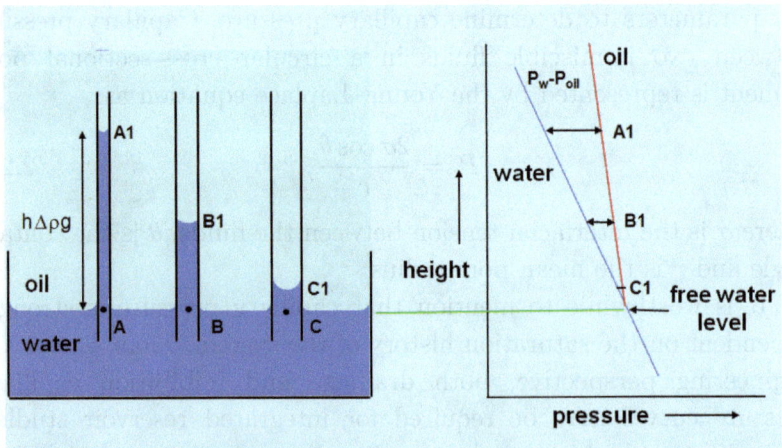

Figure 2.26. An illustration of hydrostatic equilibrium in capillary tubes.

As can be seen from Eq. (2.24) and Figure 2.26, the equilibrium height of the water column, h, is a function of the radius of the capillary tube and the wettability. If the wetting characteristics of the medium remain unaltered and the radius of the capillary tube decreases, the column height increases proportionally with the decrease in radius. If the radius of the capillary tube is kept constant and the wetting characteristic of the solid changes, then the height of the water column increases or decreases depending on the direction of the change in wettability. For example, if the system had initially been less water-wet (contact angle θ increases), then the column height would have been shorter, as the capillary forces would have become weaker.

2.6.3.2. *Characteristics of capillary pressure curve*

As discussed above, the capillary pressure curve depends not only on the fluid saturation but also on the saturation history of the system. Figure 2.27 shows a typical capillary pressure curve for an oil–brine system for drainage and imbibition cycles. It is worthwhile to mention that although in theory the non-wetting phase displacing the wetting phase is called drainage, and the wetting phase displacing

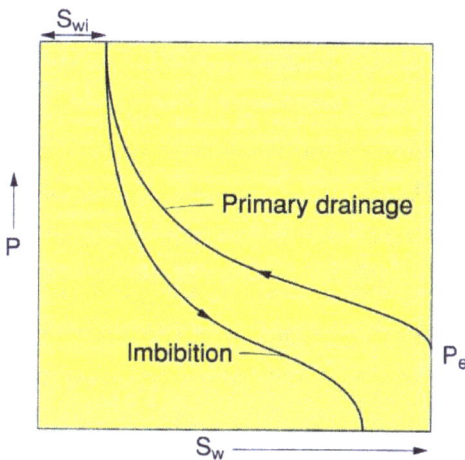

Figure 2.27. A typical capillary pressure curve for primary drainage and imbibition cycle [5].

the non-wetting phase is called the imbibition process, the industry convention is to use the drainage term for oil-displacing-water, and the imbibition term for water-displacing-oil, irrespective of the system's wettability. A minimum threshold pressure must be exceeded in order to initiate the invasion of the oil phase, which is called entry capillary pressure for primary drainage (P_e in Figure 2.27). As the pressure increases further, water saturation decreases and reaches the irreducible water saturation (or connate water saturation, S_{wc}). This point reflects the fluid distribution typically found at the time of reservoir discovery. Therefore, it is also called the initial water saturation, S_{wi} (see Figure 2.27). Once water is injected into the system after the primary oil drainage, it usually follows a different path, which is called the imbibition capillary pressure curve. This difference between drainage and imbibition capillary pressure curves is called the hysteresis effect and it has a significant impact for reservoir engineering applications. The primary drainage process represents hydrocarbon accumulation and fluid distribution at initial reservoir conditions, whereas imbibition represents the displacement performance upon water injection into the reservoir. Therefore, it is crucial to utilise the proper capillary pressure curve for a specific reservoir application.

The shape of the capillary pressure curves depends on the pore-size distribution and hence relates to the permeability of the reservoir rock. This also determines the height of the transition zone above the Free Water Level (FWL). Low permeability rocks have high capillary pressures and generally long transition zones whereas high permeability rocks have low capillary pressures and short transition zones. Figure 2.28 illustrates the impact of absolute permeability on capillary pressure curves.

It is worthwhile to mention that most carbonate reservoirs in the world (a significant proportion are located in the Middle East) show mixed-wet to oil-wet behaviour. In these reservoirs, a rather significant hysteresis effect can be observed between the primary drainage and imbibition capillary pressure curves (see Figure 2.29). Not only their wettability characteristics but also the low matrix permeabilities associated with them makes the transition zone modelling

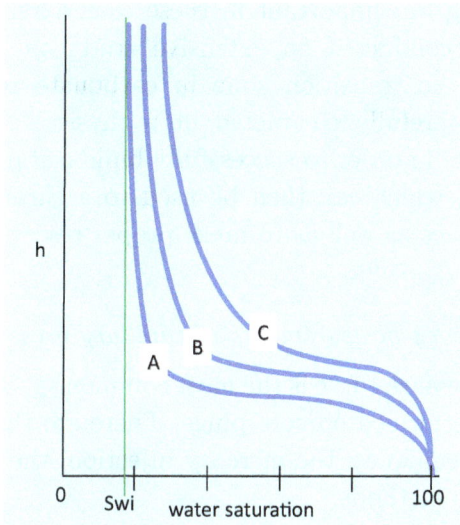

Figure 2.28. An illustration of the impact of permeability of a rock on the capillary pressure curve. A, B and C represent the capillary pressure curves of three different core samples, C being the least permeable and A being the most permeable.

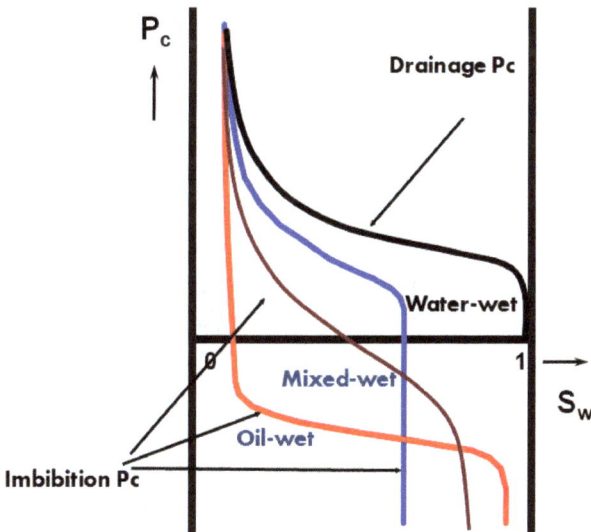

Figure 2.29. An illustration of the capillary pressure hysteresis with varying degree of wettabilities. Note that once the system becomes oil-wet, there is almost no spontaneous imbibition.

quite challenging but important in these reservoirs. Masalmeh and co-workers [20] conducted an extensive study on improved characterization of the transition zone in carbonate reservoirs. They showed that a carefully conducted petrophysical and SCAL data analysis is crucial in order to successfully build a static and dynamic reservoir model, which can then be used to accurately predict the original oil in place as well as to make proper reservoir performance predictions.

2.6.3.3. *Laboratory measurements of capillary pressure*

The capillary pressure curve is the most common special core analysis measurement performed on core plugs. There are three widely used measurement techniques: the mercury injection, the centrifuge, and the porous plate methods.

2.6.3.3.1. Mercury injection measurement

In the mercury injection technique [21], mercury is the non-wetting fluid and air is the wetting phase. The core is initially cleaned and dried. Then the mercury is injected into the clean, dry core sample by increasing the injection pressure where the volume of mercury entering into the sample is measured at each pressure step. Mercury injection is a frequently used capillary pressure measurement technique as it is relatively cheap, fast and requires relatively straightforward data interpretation. The measured data, however, need to be converted to *in situ* reservoir conditions by taking into account the differences in interfacial tension and contact angle between the rock/fluid systems used in the laboratory and that found in reservoir (see Section 2.6.3.4). The main disadvantage of mercury injection measurement is that it is only applicable for the drainage cycle; one of the other techniques should be employed for obtaining the capillary pressure curve that is representative to the imbibition cycle. A mercury injection experiment should be conducted as the last measurement since this technique is destructive and samples cannot be used for further measurements. Figure 2.30 shows a schematic of the mercury injection apparatus.

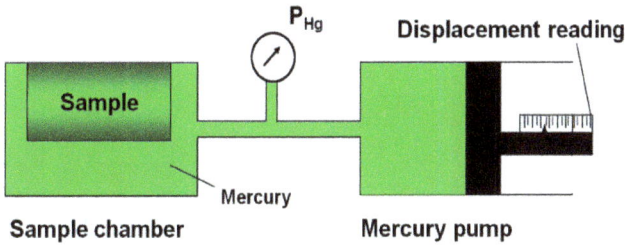

Figure 2.30. A schematic of the mercury injection apparatus.

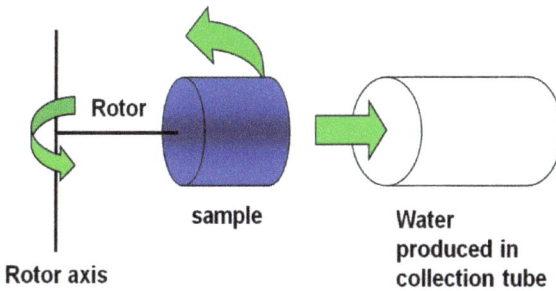

Figure 2.31. An illustration of capillary pressure measurement using the centrifuge method.

2.6.3.3.2. Centrifuge technique

Another capillary pressure measurement technique is the multi-speed centrifuge method [22] Brine- (or oil-) saturated samples are placed in a centrifuge and are spun at a series of increasing constant speeds. The rotation speed is converted into a force unit, which determines the capillary pressure. It is a relatively fast (compared to the porous plate measurement) and non-destructive technique. A complete set of capillary pressure data points can be obtained in a few days for relatively high-permeability samples and a longer time is needed for low-permeability samples. The main drawbacks of this technique are that the design of the experiment and interpretation of the data are not straightforward and numerical simulation of centrifuge experiments is generally required to derive capillary pressure data. Figure 2.31 shows the centrifuge process. For drainage experiments, brine-saturated cores are used, whereas for imbibition, oil-saturated

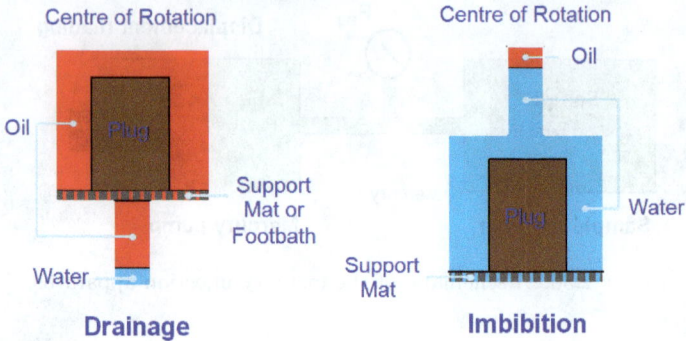

Figure 2.32.   Core is saturated with brine for drainage, and with oil for imbibition experiments (Courtesy Core Laboratories).

(saturated with oil and connate water) cores are utilised (see Figure 2.32).

2.6.3.3.3.   Porous plate method

The porous plate method is another widely used capillary pressure measurement technique, which can also be employed for resistivity index measurements (see Section 2.6.1). It is applicable for both drainage (oil-displacing-water) and imbibition (water-displacing-oil) capillary pressure measurements.

For the drainage capillary pressure measurement, a clean sample is initially saturated with the wetting fluid (brine or oil). The sample is then placed in a semi-permeable membrane, which is only permeable to the wetting phase. The gas (air or nitrogen) pressure is then increased in order for the gas to invade the core by expelling the wetting fluid through the porous plate. Once the equilibrium is reached (when no more wetting fluid production is observed), the change in the saturation is determined volumetrically. The process is repeated for the next pressure step until sufficient data points are collected. The main drawback of this technique is its being slow, since reaching pressure equilibrium at each saturation point takes a significant amount of time, which renders the technique impractical for certain field applications, especially for tight and heterogeneous carbonates. The advantage of the porous plate measurement is that

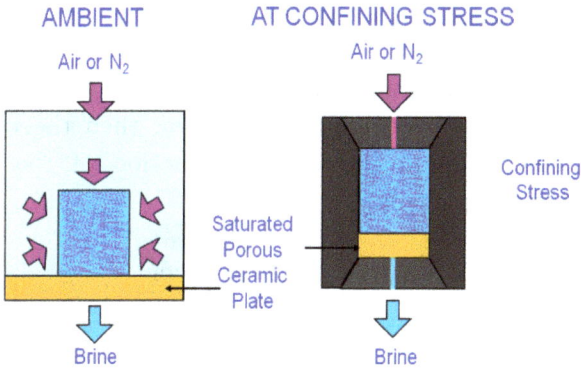

Figure 2.33. A schematic of capillary pressure measurement using the porous plate method. Note that the experiments can be conducted at either ambient conditions or at conditions where a confining stress is applied (Courtesy Core Laboratories).

it can be conducted both at ambient conditions and at the representative reservoir conditions of confining stress and temperature, as well as using reservoir fluids (see Figure 2.33).

2.6.3.4. *Laboratory to reservoir conversion*

Fluids used in laboratory measurements may not have the same physical properties as the reservoir fluids. Therefore, experimentally obtained P_c values need to be converted to equivalent reservoir conditions. The equation for laboratory to field P_c conversion is

$$P_{c,R} = \frac{\sigma_R \cos \theta_R}{\sigma_L \cos \theta_L} P_{c,L}, \tag{2.25}$$

where the subscripts R and L represent reservoir and laboratory conditions, respectively.

It is worthwhile to mention that this conversion only accounts for the changes in interfacial tension and contact angle (wettability). However, there may be further alterations due to a number of different phenomena. Stress effects in the reservoir and a wider range of pore-size distribution in the field can be considered as examples that require additional experimental measurements. Moreover, the possible field-wide variation of the wettability (i.e. contact angles), especially if the system is not strongly wetting, may require further experimental assessments.

For example, if we assume that both mercury/air and water/air systems have similar wetting characteristics, then the ratio of the capillary pressure would be solely a function of the interfacial tensions and can be calculated as follows:

$$\frac{P_c^{\text{Hg/air}}}{P_c^{\text{water/air}}} = \frac{\sigma_{\text{Hg/air}}}{\sigma_{\text{water/air}}} = \frac{480 \,\text{dyn/cm}}{70 \,\text{dyn/cm}} = 6.57. \qquad (2.26)$$

However, experience has shown that, for conversion from laboratory measurements to reservoir oil–water systems, a range of conversion factors (Eq. (2.25)) have been obtained depending on the fluids and wetting conditions. Therefore, the laboratory-field conversion (Eq. (2.25)), although useful, may not always be sufficient and further calibration, either using core measurements under representative reservoir conditions or log-derived saturations, may be needed, especially for water–oil–rock systems.

2.6.3.5. *Averaging capillary pressure data*

Capillary pressure measurements in the laboratory are performed on core plugs, which represent only a very small part of the reservoir. Therefore, attempts have been made to fit the capillary pressure data in one general curve to construct the saturation/height function for a specific reservoir. Although several different techniques were suggested in the literature, the most commonly used approach is the Leverett J-function. Leverett [23] proposed the following relationship in order to convert all the experimentally measured capillary pressure data into one universal curve:

$$J(S_{\text{w}}) = \frac{P_c}{\sigma} \sqrt{\frac{K}{\phi}}, \qquad (2.27)$$

where P_c is the capillary pressure measured in the laboratory, σ is the interfacial tension, K is the permeability and ϕ is the porosity. Note that the contact angle term ($\cos \theta$) was later added in the denominator, hence the equation becomes

$$J(S_\mathrm{w}) = \frac{P_\mathrm{c}}{\sigma \cos \theta} \sqrt{\frac{K}{\phi}}. \tag{2.28}$$

The J-function was first proposed as a means of fitting all capillary pressure into one general curve. However, there is considerable difference in the correlation between the J-function and water saturation from one formation to another. Therefore, although it may be impossible to obtain one universal curve for all the capillary pressure data, a good fit can be obtained for the data from the same rock-type of a formation.

Once the capillary pressure data are represented by a general curve, it may then be possible to convert the capillary pressure versus saturation curve into height (above free water level) versus saturation. This conversion based on the principle of capillary-gravitational force balance is

$$h = \frac{\sigma_\mathrm{R} \cos \theta_\mathrm{R}}{\sigma_\mathrm{L} \cos \theta_\mathrm{L}} \frac{P_\mathrm{c,L}}{0.433 \Delta \rho}, \tag{2.29}$$

where h is the height above the free water level (ft), σ is the interfacial tension (dyne/cm), θ is the contact angle, $P_\mathrm{c,L}$ is the capillary pressure measured in the laboratory (psi), and $\Delta \rho$ is the density difference between the two immiscible fluids, which is represented as specific gravity and hence dimensionless. 0.433 is the gravity constant, which is in psi/ft.

2.6.4. *Relative Permeability*

In Section 2.5.2 we discussed the measurement techniques of absolute permeability. We defined absolute permeability as the ability of a formation to transmit a fluid. However, if the system is saturated with more than one fluid, the concept of permeability needs be extended in order to properly account for the conductivity of each fluid flowing in the pore space. The concept of relative permeability allows the establishment of a relationship between the saturation and the conductivity of a fluid. Therefore, to be able to calculate the flow rate of a fluid using Darcy's equation, the absolute permeability term is replaced by the effective permeability, which can be represented as

$$k_{\text{eff}} = k * k_{\text{r}}, \tag{2.30}$$

where k is the absolute permeability of the formation and k_{r} is the relative permeability of the system to a specific fluid.

Unlike absolute permeability, which is a pure reservoir rock property, relative permeability is controlled by the following factors: wettability, pore geometry, interfacial tension, fluid saturation and saturation history. Since it is a function of several different parameters, the measurement of relative permeability is not as straightforward as that of absolute permeability. Figure 2.34 illustrates typical oil and water relative permeability curves for a water–oil system.

An accurate estimate of relative permeability data is essential for almost all reservoir-engineering applications. In order to forecast the performance of a waterflood or to design an enhanced oil recovery project, one needs to input the constitutive relationship between the relative permeability and the phase saturation into the simulation

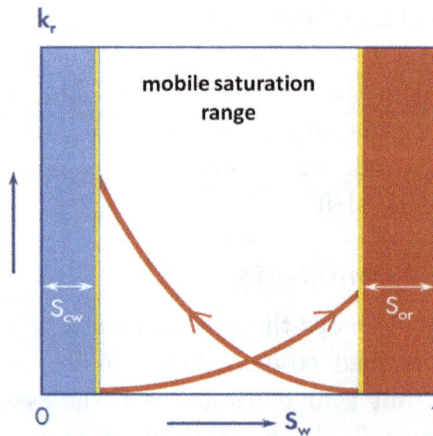

Figure 2.34. An illustration of a typical relative permeability versus saturation relationship for a water–oil system. Note that the region shown by white colour represents the saturation range in which both oil and water phases are flowing (mobile). S_{cw} is the immobile connate-water saturation, and S_{or} is the immobile residual oil saturation.

model. As this is influenced by many physical parameters, every effort should be made to mimic the reservoir conditions as closely as possible while performing a relative permeability measurement. However, conducting experiments at reservoir conditions is often complex and expensive. Although it may not be possible to perform reservoir-condition measurements in many cases, it would be necessary to use native-state (or restored-state) cores during these measurements to mimic reservoir wettability.

2.6.4.1. *Laboratory measurements of relative permeability*

Relative permeability is one of the most important multi-phase flow parameters that can be measured in the laboratory. As the flow rate calculations and hence reservoir performance predictions have direct dependence on the relative permeability data, it is essential to make the measurement as accurate as possible. There are three widely used measurement methods: steady-state, unsteady-state and centrifuge. They are described below.

2.6.4.1.1. Steady-state method

In a steady-state relative permeability measurement, two immiscible fluids are simultaneously injected into a small core plug at a predetermined flow ratio. The injection continues until a steady-state condition is reached (when the produced fluid ratio is equal to the injected fluid ratio and pressure drop becomes stable). Once the steady-state condition is obtained, the existing fluid saturations inside the core sample are considered to be stable. Therefore, they can be measured through a volumetric (or gravimetric) material balance or by employing an X-ray CT scanner. Once the saturations are measured, relative permeability of both phases can be calculated at these saturation points using Darcy's equation. The fractional flow ratio between the injected fluids is then altered. If a drainage relative permeability measurement is conducted, then the flow rate of the non-wetting phase is increased, whereas if imbibition relative permeability is measured, then the flow rate of the wetting-phase is increased. The injection then continues with the new fractional flow ratio until a new steady-state condition is reached. The new phase

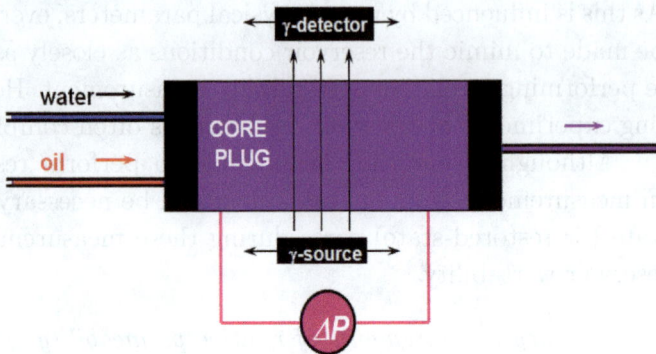

Figure 2.35. Experimental apparatus for steady-state relative permeability measurement.

saturations are then measured, and the same procedure is repeated for a number of different fractional flow ratios until a complete relative permeability curve is obtained.

The steady-state measurements provide experimental data in a wide range of saturation values. However, the reliability of the measurements in the low saturation region is usually questionable due to the rate effects associated with viscous instabilities. Therefore, experimental data are useful particularly for the middle saturation range. Although it is a recommended measurement technique, steady-state measurements often take too long to complete. Reaching a steady-state condition for each measurement point is a lengthy process. Experiments usually take up to two weeks depending on the properties of the sample and number of data points measured. Figure 2.35 illustrates a schematic of the experimental apparatus for the steady-state measurements.

2.6.4.1.2. Unsteady-state method

The unsteady-state relative permeability measurement involves the displacement of a fluid by another fluid through a relatively homogenous core sample. During the experiment, it is sufficient to monitor the cumulative volume of the displacing and displaced fluids as a function of time. Analytical techniques are then used to calculate the phase relative permeabilities by assuming a constant pressure

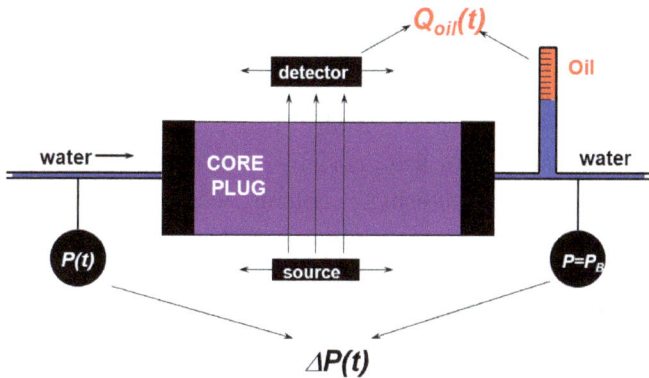

Figure 2.36. Experimental apparatus for unsteady-state relative permeability measurement.

differential across the core sample, which is assumed to be the same for each phase, and by ignoring the capillary pressure and capillary end effects (this will be discussed in more detail in the next section). The method proposed by Welge [24], which is dependent upon the frontal advance fluid-flow concept, is applied to calculate the relative permeability ratio. Standard interpretation techniques such as the JBN method [25] can then be used to determine the individual phase relative permeabilities.

Although the unsteady-state measurement is quick and hence widely used, the interpretation of the data is not straightforward. Furthermore, the analytical techniques used to calculate the relative permeabilities are based on two assumptions (sample homogeneity and displacement stability). Therefore, the reliability of the unsteady-state measurements is questionable in many circumstances. Figure 2.36 illustrates a schematic of the experimental apparatus for the unsteady-state measurements.

2.6.4.1.3.   Centrifuge method

In a centrifuge experiment, samples are first placed in centrifuge core holders; then the invading phase displaces the fluid from the samples. The cumulative volume of the displaced fluid is monitored with respect to time. Analytical techniques [26] are used to calculate the relative permeability curves. However, the recommended approach is to history-match the production data using computer simulation in order to obtain the relative permeabilities [2].

The centrifuge technique has some advantages over alternative methods. First, it is substantially faster than the steady-state technique. Displacements are gravity-stable and hence they are not subject to the viscous fingering problems, which sometimes interfere with the unsteady-state measurements. As the relative permeabilities can be measured to very small values, the centrifuge method is often considered to be the best technique for determining endpoint saturations such as the residual oil saturation, S_{or}. On the other hand, the centrifuge method, like other techniques, is subject to capillary end-effect problems. Furthermore, this method does not provide the relative permeability curves for both phases on the same cycle. In a centrifuge experiment, only the relative permeability to the invaded phase is determined. Therefore, for an imbibition process, the non-wetting phase relative permeability is calculated, whereas for a drainage process, the wetting-phase relative permeability is calculated. Figure 2.32 illustrates the centrifuge process.

These are the most widely used relative permeability measurement techniques. If available, it would be useful to validate the measured experimental data by comparing the results obtained by other measurement techniques. It is also often recommended to combine the data sets calculated by different measurement techniques to establish a full relative permeability curve. Ideally, steady-state measurements for the middle-saturation range and centrifuge measurements for low-saturation region can be combined to obtain a full relative permeability curve.

One of the recent advances in relative permeability measurements is employing core-flood simulators in order to history-match the measured production data. As already mentioned, analytical techniques used to interpret the measured data are based on some assumptions, which may or may not be valid, such as ignoring the capillary pressure and the end effects. However, core-flood simulations do not suffer from these shortcomings. Numerical models are capable of accounting for both capillary pressure and gravity effects and hence they honour the actual flow processes and boundary conditions.

2.6.4.2. *Extrapolating measured relative permeability data*

It may often be the case that measured relative permeability data points are insufficient for the purpose of reservoir simulation studies. For example, unsteady-state measurement provides data only after the breakthrough of the invading phase. However, simultaneous flow of the multiple phases does occur in the reservoir even if the injected fluid does not reach the production well. Therefore, it is often necessary to extrapolate the laboratory-measured relative permeability data to a wider range of saturation regions.

Corey [27] suggested a model that relates the end-point relative permeability to the saturation of the phase. For a water–oil system, the connate water saturation, S_{wc}, the residual oil saturation, S_{or}, the end-point oil and water relative permeabilities and the constants, which are known as the Corey exponents, are the only parameters necessary to establish the full oil and water relative permeability curves. The Corey model is widely used in reservoir engineering applications and can be represented as

$$K_{rw} = K_{rw}(S_{or}) \left(\frac{S_w - S_{wc}}{1 - S_{wc} - S_{or}} \right)^{N_w} \tag{2.31}$$

$$K_{ro} = K_{ro}(S_{wc}) \left(\frac{1 - S_w - S_{or}}{1 - S_{wc} - S_{or}} \right)^{N_o}, \tag{2.32}$$

where N_w and N_o are the Corey exponents for water and oil, respectively, which depend on the permeability and the pore structure of the rock and the wettability of the system; $K_{rw}(S_{or})$ is the end-point relative permeability to water, which is measured when the system is saturated with water and the residual oil; $K_{ro}(S_{wc})$ is the end-point relative permeability to oil, which is measured when the system is saturated with oil and the connate water.

2.6.4.3. *Factors affecting relative permeability measurements*

Relative permeability is influenced by many physical parameters, such as wettability, pore geometry and saturation history. Therefore, extra care should be taken while planning and executing the relative

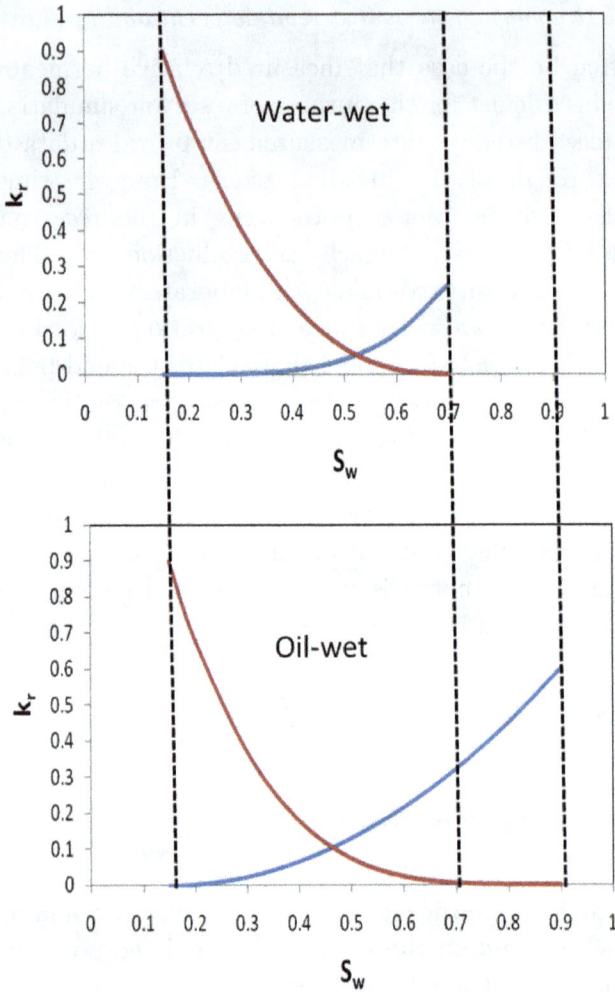

Figure 2.37. Effect of wettability on relative permeability. Note that an oil-wet core shows higher water productivity as well as lower residual oil saturation.

permeability measurements. It would be the best to perform the experiments at the wetting conditions identical to the reservoir wettability. Figure 2.37 illustrates the impact of wettability on relative permeability. Compared to a water-wet system, an oil-wet (or a mixed-wet) rock usually shows higher water productivity and lower residual oil saturation. Therefore, even if the test is conducted at identical reservoir conditions, it is crucial to use the right fluid combination with the representative rock wettability.

Another important consideration is the saturation history. The wettability of a system depends on the displacement cycle. Contact angles, and hence the wettability characteristics, are believed to be different for the drainage and the imbibition processes. The term "advancing contact angle" is used to represent the contact angle for an imbibition process and "receding contact angle" is used for a drainage cycle. The advancing contact angle is often larger than the receding contact angle. The difference between the advancing and the receding contact angles is called the contact angle hysteresis and may be significant depending on the surface roughness and the surface contamination [28, 29]. Therefore, it is essential to conduct relative permeability experiments that closely mimic the saturation history of the reservoir.

Even if the tests are conducted with the right fluid combination and at the identical reservoir conditions, there may still be shortcomings due to the nature of the measurement techniques. One important issue in core-flood experiments is the capillary end effect that arises from the discontinuity of the wetting phase saturation at the outflow face. In a permeable medium, the capillary forces act uniformly in all directions. However, this is not the case at the outlet of a core sample. When the flowing phases are discharged into an open region under atmospheric pressure, a net capillary force persists in the sample, which does not allow the wetting phase to leave the sample. Therefore an accumulation of the wetting phase occurs at the outflow face, which creates a saturation gradient along the sample and disturbs the relative permeability measurements [30]. Figure 2.38 illustrates a typical saturation profile obtained in a core-flood experiment where the wetting phase is denoted by the non-shaded region.

Capillary end effects most commonly appear in cases where the non-wetting phase is displacing the wetting phase, such as oil displacing water in a water-wet rock. Oil-displacing-water experiments are specifically important, as they establish the end-point oil relative permeability, K_{ro}, at connate water saturation, S_{wc}, which is the starting point of a waterflood process [31].

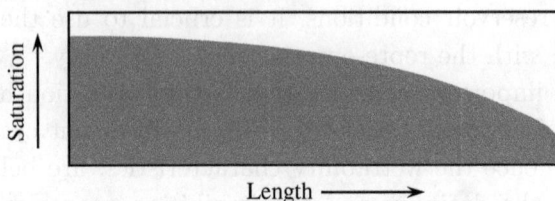

Figure 2.38. A schematic illustration of the capillary end effect. The non-wetting phase saturation is shaded in grey while the wetting phase is not shaded. Note the wetting phase accumulation at the outflow face (right hand side).

There are several techniques that are widely used to minimise the end effects due to the capillary forces. One approach is to add small end pieces to the test sample in order to establish a capillary continuity, which may reduce the wetting-phase accumulation at the outlet. Another technique for minimising the capillary end effects is to conduct the experiments with high flowrates. As the impact of capillary forces reduces with the increasing flowrates, the amount of wetting phase retained at the outflow face would be significantly lower. Another approach is to use a longer core plug. However, care should be taken while choosing the plug since the sample may become more heterogeneous once the size increases. If *in situ* saturation measurement is employed during the core-flood experiment, a common approach for reducing the end effects is to measure the saturation and pressure drop via pressure taps placed along the sample length by ignoring a small fraction of the plug at the inflow and outflow faces of the sample. Numerical simulation of displacement experiments can be used to history-match experiments and derive representative relative permeability and capillary pressure data if all the necessary data, including *in situ* fluid saturation distribution, are available.

2.6.4.4. *Three-phase relative permeabilities*

With the increasing maturity of the world's producing fields, improved recovery of these resources to boost additional production is becoming increasingly important. Many enhanced oil recovery projects involve the simultaneous flow of three phases: water, oil and gas. In particular, the use of CO_2 injection into mature oil fields is likely to become more common not only to enhance the oil recovery but also for the storage of the CO_2. In order to design such a project and make performance predictions, one needs to input the relative permeabilities of all three phases into the simulation model.

However, experimental measurements of three-phase relative permeabilities are tedious and time-consuming. In addition to the measurement of the saturations, the pressure drops and the fluxes for three flowing phases, there are an infinite number of different displacement paths. Thus, it is impractical to measure relative permeability for all possible three-phase displacement scenarios that may occur in a reservoir. As a result, the universal practice in reservoir simulation studies is to estimate the three-phase relative permeabilities using empirical correlations, which rely on two-phase relative permeability data.

The most widely used empirical correlations to predict three-phase relative permeabilities are the two models developed by Stone in 1970 and 1973 respectively, which are known as Stone I [32] and Stone II [33], as well as the so-called saturation-weighted interpolation which was developed in 1988 [34]. Although a number of other methods have been proposed, these three are the most popular ones and they are implemented in most commercial reservoir simulators.

Empirical models normally estimate the three-phase relative permeabilities by employing the measured two-phase relative permeability data. For instance, the oil relative permeability, K_{ro}, is estimated as a function of two different oil relative permeabilities, which are measured in a water–oil experiment, $K_{ro(w)}$, and in a gas–oil experiment, $K_{ro(g)}$. However, pore occupancies and displacement orders can be substantially different in three-phase flow, and hence may not be represented accurately by two-phase experiments [35]. Therefore, although they are widely used in many reservoir engineering applications, the accuracy and reliability of the empirical equations are questionable, especially for the systems with non-homogeneous wettability. Moreover, recent studies suggest that currently used empirical models may fail to predict the three-phase relative permeabilities accurately [36, 37].

It has been shown that the relative permeability to the gas phase may show significant hysteresis for cyclic injection processes such as Water-Alternating-Gas (WAG) floods [37]. Therefore, it is essential to accurately predict not only the three-phase relative permeabilities but also their dependence on the saturation history. A recent study, conducted on the performance estimation for gas-based Enhanced Oil Recovery (EOR) processes, shows that cyclic injection processes cannot be accurately modelled unless the relative permeability hysteresis is accounted for [38].

2.7. Recent Developments in Core Analysis

Although most RCAL and SCAL measurement techniques remain essentially the same, there has been a significant improvement in the instrumentation. The enhancements in the imaging techniques and the increase in computational power provide a great variety of help for laboratory measurements. Nuclear Magnetic Resonance (NMR) and micro-CT scanning are two such technologies, which are becoming increasingly popular for evaluating different petrophysical parameters. Pore-scale modelling is yet another technique, which can be used as a complementary tool for core analysis laboratory measurements.

2.7.1. *Nuclear Magnetic Resonance*

NMR was first discovered in 1945 and has been used in petrophysical data analysis for more than five decades now. NMR technology is applied both in logging applications via Magnetic Resonance Image Logs (MRIL) and in core analysis via NMR spectrometers.

NMR spectrometers use the Radio Frequency (RF) resonance of protons in a magnetic field to determine the longitudinal (spin–lattice) relaxation time, T_1, and transverse (spin–spin) relaxation time, T_2. The measurement of T_1 and T_2 requires different pulse sequences. The common practice is to use T_2 because in most cases T_2 is smaller than T_1 and hence can be determined faster. The

NMR relaxation is attributed to the hydrogen relaxation originating from the fluids residing in the pore space. The relaxation of the hydrogen protons is governed by the pore structure of its local environment. Since the pore space of a core sample consists of billions of "pores" with different shapes and sizes, a wide range of T_2 values are measured and plotted as a histogram with respect to the pore size distribution.

Historically, T_2 measurements have been used as a good indicator for pore-size distribution, porosity and fluid saturations. Although, this is still the main goal of NMR applications, there have been substantial developments in the quantitative estimation of the wettability indices [39] and in the prediction of the permeability through its pore-size dependence, specifically for systems with fairly uniform grain-size distribution where a thorough relationship between the pore size and permeability can be established [40].

2.7.2. *Pore-Scale Network Modelling*

Pore-scale network models are numerical simulators that are used to model the displacement mechanisms at a few-micron resolution by defining the pore space of a rock with a network of pore bodies and pore throats. Modelling the flow behaviour using network models was pioneered by Fatt in 1956 [41] Since then, many authors have developed pore-scale network models to simulate the single- and multi-phase fluid flow through porous media. However, further advances in network modelling were only possible after the late 1990s when computer processing power became more readily available.

The two main challenges associated with the network modelling are an accurate characterisation of the pore space of the rock and a proper definition of the system's wettability and its variations. There have been significant attempts to address these issues in the last two decades. Øren and co-workers [42] introduced the so-called process-based network generation algorithm, where a thin-section image is utilised to construct a 3D network model by simulating the geological processes such as sedimentation, compaction and diagenesis. It was suggested that although the technique is successful in some sandstones, this might not be the case in carbonates due to the complexities associated with the modelling of complex pore structures caused by diagenesis. Attempts have also been made to address the issues related to the wettability characterization. Valvatne and Blunt [43] assigned a range of contact angles to the pores and throats, as well as accounting for the hysteresis effect by altering the contact angles for the drainage and imbibition cycles in order to properly characterise the wettability of the rock.

With high-resolution micro-CT scanning technology becoming more readily available in oil industry applications, substantial steps have been taken in pore-space imaging and characterization [44]. With the increase in computational power, it may also be possible to conduct simulations on the direct 3D images by avoiding the bottlenecks of generating a network of pore and throat elements [45]. However, a proper definition of the wettability and its variation in time and space remain as important challenges, which limit the predictive capabilities of the pore-scale modelling tools.

Overall, pore-scale modelling that combines an accurate description of the pore space with a detailed analysis of pore-scale displacement physics is an excellent tool for understanding multi-phase flow in porous media. From a practical perspective, it can be used as a complementary tool for core analysis laboratory measurements and also has the potential to make predictions for situations that are difficult to study experimentally, such as three-phase flow experiments [37, 46, 47].

2.7.3. *Core Analysis for Unconventional Resources*

With the increasing production from shales and other tight rocks, there is an interest in applying core analysis methods to these rock types in order to understand more about the rock properties and flow behaviour. Unfortunately it is in most cases impossible to investigate shales with existing conventional core analysis methods. This is because shales have low porosities (typically less than 10%) and permeability (in the nano-Darcy range; a factor 10^6 smaller than a MD rock), and they retain fluids through physical processes not normally seen in conventional reservoir rocks (such as adsorption).

New analysis methods are therefore applied to tight rocks. For example, absolute permeability can no longer be measured with traditional methods where the steady-state flow of a liquid through the sample is achieved and a differential pressure measured. Shales are too tight for this and it is not feasible to wait for steady-state conditions to be attained in the laboratory. Instead, transient methods are used where, for example, a pressure pulse is applied to one end of the sample and the delay in measuring the pulse at the other end is recorded [48]. Handwerger and co-workers [49] give an overview of some of those techniques which are introduced to improve core analysis for shale reservoirs.

Acknowledgements

The authors would like to thank Shell management for permission to publish this work and Shell Learning Centre for making many of the figures and illustrations used in this work available. They would also like to thank Fons Marcelis of Shell International Exploration & Production and Shehadeh Masalmeh of Shell Abu Dhabi for the technical review.

References

[1] Archie, G.E. (1950). Introduction to Petrophysics of Reservoir Rocks, *AAPG Bulletin*, **34**(5), 943–961.

[2] Boyle, K., Jing, X.D. and Worthington, P.F. (2000). Petrophysics — Modern Petroleum Technology, in R.A. Dawe (Ed.), *The Institute of Petroleum Publication*, John Wiley & Sons Ltd., West Sussex, England.

[3] Worthington, P.F. and Longeron, D. (1991). *Advances in Core Evaluation II: Reservoir Appraisal*, Gordon and Breach Science Publishers, Philadelphia, PA, USA.

[4] Amyx, J.W., Bass, D. and Whiting, R.L. (1960). *Petroleum Reservoir Engineering: Physical Properties*, McGraw-Hill Inc., New York City, NY, USA.

[5] Yuan, H.H. and Schipper, B.A. (1996). Core Analysis Manual, *Shell E&P Report*.

[6] Skopec, R.A. and McLeod, G. (1996). Recent Advances in Coring Technology: New Techniques to Enhance Reservoir Evaluation and

Improve Coring Economics, *Journal of Canadian Petroleum Technology*, **36**(11), 22–29.

[7] Park, A. (1983). Improved Oil Saturation Data Using Sponge Core Barrel, *SPE Production Operation Symposium*, 27 February–1 March, Oklahoma City, OK, USA.

[8] Skopec, R.A. (1994). Proper Coring and Wellsite Core Handling Procedures: The First Step Toward Reliable Core Analysis, *Journal of Petroleum Technology*, **46**(4), 280.

[9] Okkerman, J.A. and van Geuns, L.J. (1993). Core Handling Manual, *Shell E&P Report*.

[10] Klinkenberg, L.J. (1941). The Permeability of Porous Media Liquids and Gases, in *API Drilling and Production Practice*, American Petroleum Institute, Washington, DC, pp. 200–213.

[11] Unalmiser, S. and Funk, J.J. (1998). Engineering Core Analysis, *Journal of Petroleum Technology*, **50**(4), 106–114.

[12] Forchheimer, P. (1901). Wasserbewegung durch Boden, *Ver Deutsch Ing.*, **45**, 1782–1788.

[13] Jing, X.D., Archer, J.S. and Daltaban, T.S. (1992). Laboratory Study of the Electrical and Hydraulic Properties of Rock Under Simulated Reservoir Conditions, *Marine and Petroleum Geology*, **9**(2), 115–127.

[14] Archie, G.E. (1942). The Electrical Resistivity Log as an Aid in Determining Some Reservoir Characteristics, *Petroleum Transactions of AIME*, **146**, 54–62.

[15] Waxman, M.H. and Smits, L.J.M. (1968). Electrical Conductivities in Oil Bearing Shaly Sands, *Society of Petroleum of Engineers Journal*, **8**, 213–225.

[16] Juhasz, I. (1981). Normalised Qv — the Key to Shaly Sand Evaluation Using Waxman–Smits Equation in the Absence of Core Data, *SPWLA 22nd Annual Logging Symposium*, 23–26 June, Mexico City, Mexico.

[17] Craig, F.F. (1971). *The Reservoir Engineering Aspects of Waterflooding*, Monograph Series, SPE, Richardson, Texas, USA.

[18] Anderson W.G. (1986). Wettability Literature Survey — Part 1: Rock/Oil/Brine Interactions and the Effects of Core Handling on Wettability, *Journal of Petroleum Technology*, **38**(10), 1125–1143.

[19] Anderson, G. (1975). *Coring and Core Analysis Handbook*, Petroleum Publishing Co. Tulsa, Oklahoma.

[20] Masalmeh, S.K., Abu Shiekah, I. and Jing, X.D. (2007). Improved Characterization and Modeling of Capillary Transition Zones in Carbonate Reservoirs, *SPE Reservoir Evaluation and Engineering*, **10**(2), 191–204.

[21] Purcell, W.R. (1949). Capillary Pressure — Their Measurements Using Mercury and the Calculation of Permeability Therefrom, *Petroleum Transactions of AIME*, **186**, 39–48.

[22] Slobod, R.L., Chambers, A. and Prehn Jr., W.L. (1951). Use of Centrifuge for Determining Connate Water, Residual Oil, and Capillary Pressure Curves of Small Core Samples, *Petroleum Transactions of AIME*, **192**, 127–134.

[23] Leverett, M.C. (1941). Capillary Behavior in Porous Solid, *Petroleum Transactions of AIME*, **142**, 151–169.

[24] Welge, H.J. (1952). A Simplified Method for Computing Oil Recovery by Gas or Water Drive, *Petroleum Transactions of AIME*, **195**, 91–98.

[25] Johnson, E.F., Bossler, D.P. and Naumann, V.O. (1959). Calculation of Relative Permeability from Displacement Experiments, *Petroleum Transactions of AIME*, **216**, 370–374.

[26] Hagoort, J. (1980). Oil Recovery by Gravity Drainage, *Society of Petroleum of Engineers Journal*, **20**, 139–150.

[27] Corey, A. (1954). The Interrelation Between Gas and Oil Relative Permeabilities, *Producers Monthly*, **19**(1), 38–41.

[28] Morrow, N.R. (1975). Effects of Surface Roughness on Contact Angle with Special Reference to Petroleum Recovery, *Journal of Canadian Petroleum Technology*, **14**, 42–53.

[29] Dullien, F.A.L. (1992). *Porous Media: Fluid Transport and Pore Structure*, 2nd edition, Academic Press, San Diego, CA, USA.

[30] Honarpour, M., Koederitz, L. and Harvey, A.H. (1986). *Relative Permeability of Petroleum Reservoirs*, CRC Press Inc., Boca Raton, FL, USA.

[31] Huang, D.D. and Honarpour, M. (1998). Capillary End Effects in Coreflood Calculations, *Journal of Petroleum Science and Engineering*, **19**, 103–117.

[32] Stone, H.L. (1970). Probability Model for Estimating Three-Phase Relative Permeability, *Journal of Petroleum Technology*, **22**(2), 241–218.

[33] Stone, H.L. (1973). Estimation of Three-Phase and Residual Oil Data, *Journal of Canadian Petroleum Technology*, **12**(4), 53–61.

[34] Baker, L.E. (1988). Three-Phase Relative Permeability Correlations, *SPE/DOE EOR Symposium*, 17–20 April, Tulsa, OK, USA.

[35] Blunt, M.J. (2000). An Empirical Model for Three-Phase Relative Permeability, *Society of Petroleum Engineers Journal*, **5**(4), 435–445.

[36] Spiteri, E.J. and Juanes, R. (2004). Impact of Relative Permeability Hysteresis on Numerical Simulation of WAG Injection, *SPE Annual Technical Conference and Exhibition*, 26–29 September, Houston, TX, USA.

[37] Suicmez, V.S., Piri, M. and Blunt, M.J. (2007). Pore Scale Simulation of Water Alternate Gas Injection, *Transport in Porous Media*, **66**(3), 259–286.

[38] Masalmeh, S.K. and Wei, L. (2010). Impact of Relative Permeability Hysteresis, IFT dependent and Three Phase Models on the Performance of Gas Based EOR Processes, *SPE Abu Dhabi International Petroleum Exhibition & Conference*, 1–4 November, Abu Dhabi, UAE.

[39] Looyestijn, W.J. and Hofman, J.P. (2006). Wettability-Index Determination by Nuclear Magnetic Resonance, *SPE Reservoir Evaluation and Engineering*, 9(2), 146–153.

[40] Fleury, M., Deflandre, F. and Godefroy, S. (2001). Validity of Permeability Prediction from NMR Measurements, *C. R. Acad. Sci. Paris, Chimie/Chemistry*, 4, 869–872.

[41] Fatt, I. (1956). The Network Model of Porous Media I. Capillary Pressure Characteristics, *Transactions of American Institute of Mining, Metallurgical, and Petroleum Engineers*, **207**, 144–159.

[42] Øren, P.E., Bakke, S. and Arntzen, O.J. (1998). Extending Predictive Capabilities to Network Models, *Society of Petroleum Engineers Journal*, 3(4), 324–326.

[43] Valvatne, P.H. and Blunt, M. (2004). Predictive Pore-scale Modeling of Two-phase Flow in Mixed Wet Media, *Water Resources Research*, **40**, 1–21.

[44] Arns, J.Y., Sheppard, A.P., Arns, C.H., Knackstedt, M.A., Yelkhovsky, A. and Pinczewski, W.V. (2007). Pore Level Validation of Representative Pore Networks Obtained from Micro-CT Images, *21st International Symposium of the Society of Core Analysts*, 10–14 September, Calgary, AB, Canada.

[45] Ovaysi, S. and Piri, M. (2010). Direct Pore-level Modeling of Incompressible Fluid Flow in Porous Media, *Journal of Computational Physics*, **229**(19), 7456–7476.

[46] Piri, M. and Blunt, M.J. (2005). Three-Dimensional Mixed-Wet Random Pore-Scale Network Model of Two- and Three-Phase Flow in Porous Media. I. Model Description, *Physical Review*, **71**(2), 026301.

[47] Piri, M. and Blunt, M.J. (2005). Three-Dimensional Mixed-Wet Random Pore-Scale Network Model of Two- and Three-Phase Flow in Porous Media. II. Results, *Physical Review*, **71**(2), 026302.

[48] Wang, J. and Knabe, R.J. (2010). Permeability Characterization on Tight Gas Samples Using Pore Pressure Oscillation Method, *24th International Symposium of the Society of Core Analysts*, 4–7 October, Halifax, Nova Scotia, Canada.

[49] Handwerger, D.A., Suarez-Rivera, R., Vaughn, K.I. and Keller, J.F. (2012). Methods Improve Shale Core Analysis, *The American Oil & Gas Reporter*, December.

Chapter 3

Production Logging

Olivier Allain

KAPPA, France

3.1. Introduction

This chapter is extracted from the KAPPA Dynamic Data Analysis (DDA) book. The interpretation workflow and options described herein reflect the workflow implemented in the KAPPA Production Logging module Emeraude, and represent what a modern interpretation software should offer.

The first production logs were temperature surveys run in the 1930s. The first spinner-based flowmeters appeared in the 1940s. They were complemented by fluid density and capacitance tools in the 1950s. Together with some Pressure–Volume–Temperature (PVT) correlations and flow models, these were the elements required to perform what we will call today a "classical" multi-phase interpretation.

The first probe tools were introduced in the 1980s. Although they acquired local measurements at discrete points across the borehole cross-section, the initial objective was to reduce the local measurements to a normal pipe averaged value. In the late 1990s, these probes were packaged into complex tools that began to measure the velocity and holdup distributions across the pipe in order to address the challenges of understanding the flow in complex, horizontal and near-horizontal wells.

Production log interpretation was too long overlooked by the industry. It is not part of a standard formation in Petroleum

Figure 3.1. An example of tool string.

Engineering departments. To our knowledge, only Imperial College in the UK has an education module dedicated to PL. With a few noticeable exceptions, most oil companies used to consider that PL was a simple process of measuring downhole rates. Specialized interpretation software was developed and used by service companies to process the data, either on site or in a computing centre. This situation changed in the mid-1990s, when commercial PL software became more accessible within the industry, and oil companies began to realize that PL interpretation was, indeed, an interpretation process and not just data processing.

PL is an in-well logging operation designed to describe the nature and the behaviour of fluids in or around the borehole, during either production or injection. We want to know, at a given time, phase by phase and zone by zone, how much fluid is coming out of or going into the formation. To do this, the service company engineer runs a string of dedicated tools (see Figure 3.1).

PL may be run for different purposes: monitoring and controlling the reservoir, analysing dynamic well performance, assessing the productivity or injectivity of individual zones, diagnosing well problems and monitoring the results of a well operation (stimulation, completion, etc.). In some companies, the definition of PL extends up to what we call cased hole logging, including other logs such as Cement Bond Logs (CBL), Pulse Neutron capture Logs (PNL), Carbon/Oxygen logs (C/O), corrosion logs, radioactive tracer logs and noise logs. In this chapter, we will focus on PL *per se* and explain the main methods to interpret classical and multiple probe production logs.

3.2. What Production Logging is Used for

PL can be used at all stages of the well life, during the natural production of the well, at the time of secondary or tertiary recovery, or when using injection wells.

It is the authors' recommendation that PL be run at early stage of the life of the well, in order to establish a baseline that will be used later when things go wrong. This is generally done when PL is identified as a necessary reservoir-engineering tool (for example, in extremely layered formations), otherwise it will be a hard sell to management.

Too often, production logs are run when "something" has gone wrong.

At the end of a successful interpretation, the PL engineer will have a good understanding of the phase-by-phase zonal contributions. When Multiple Probe Tools (MPT) are used, this will be complemented by image tracks describing the flow distribution (Figure 3.2).

PL allows the qualification and/or the quantification of a certain number of operational issues: formation cross-flow, channelling of undesired phases via a poor cement, gas and water coning, casing leaks, corrosion, non-flowing perforations, etc. (Figure 3.3).

PL can also be used to identify fracture production and early gas or water breakthrough via high permeability layers (Figure 3.4). During shut-ins, the movement of a water or oil column can also be identified (Figure 3.5).

3.3.   Classical Production Logging Tools

The schematic of a typical PL job is shown in Figure 3.6(a). During a stabilised flow period (production, injection or shut-in), the PL tool string, hanging on a cable controlled by a logging unit, is run up and down in front of the contributing zones at different speeds. There are also static transient surveys, referred to as "stations", where the tool is immobilised at different depths. From these runs, the PL interpretation engineer will calibrate the tools, then calculate a flow profile.

PL tool strings may be run with surface read-out, using a monoconductor cable or on slickline. Surface read-out allows real-time quality control of the data at surface. The logging program can then be adjusted depending on the results. The slickline option is less expensive and easier to run; however, it is a blind process, as data are stored in the tool memory. There is no way to check the

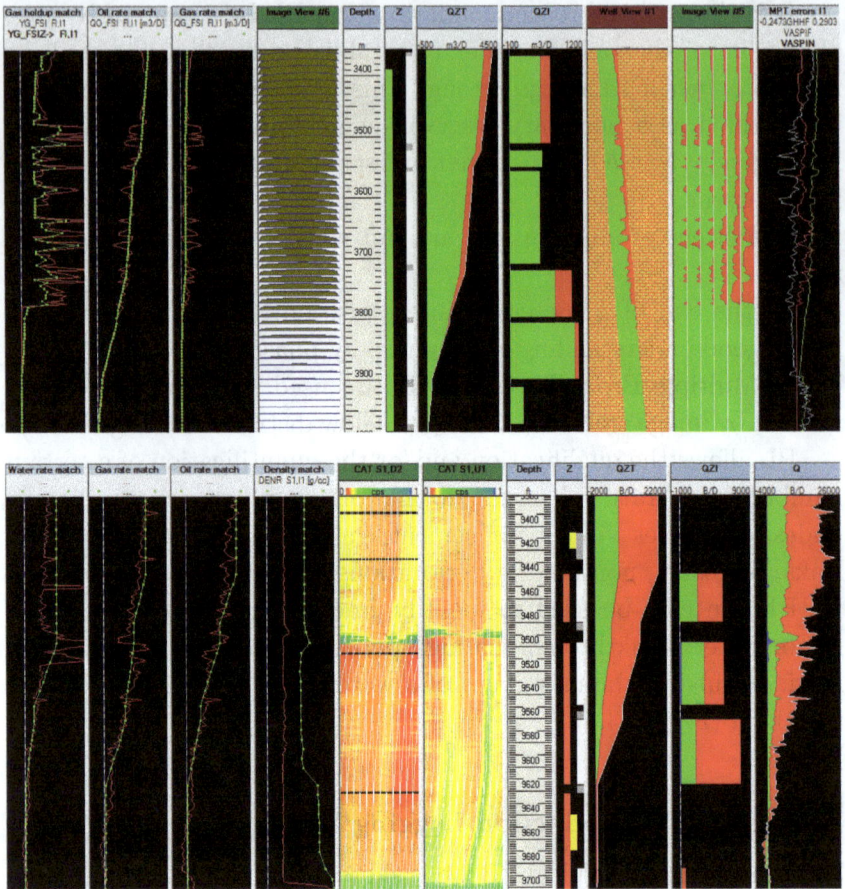

Figure 3.2. MPT interpretations (FSI top and MAPS bottom).

data before the string is pulled out. Power requirement also limits the range of tools that can be run on slickline.

A typical, classical tool string is shown in Figure 3.6(b). Sensors are not located at the same measure point of the string. As measurements are presented versus depth, they are not synchronised for a given depth, and this may become a serious issue when flow conditions are not stabilised. In some cases, one may have to stop the tool and check for the transients, in order to discriminate noisy tools and real effects. Conversely, when running stationary surveys, tools are not recorded at the same depth. This issue of time versus

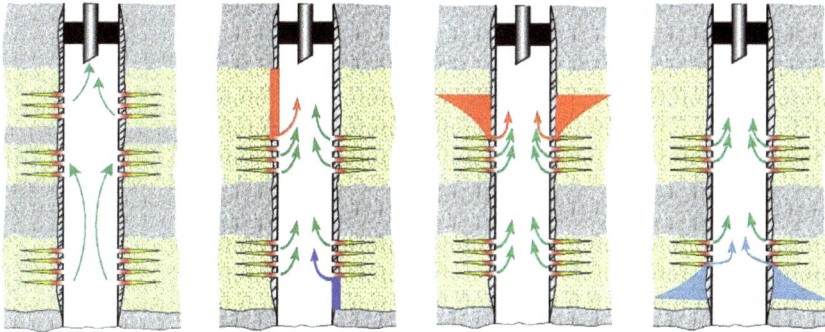

Figure 3.3. Examples of usage: cross-flow, channelling, gas coning and water coning.

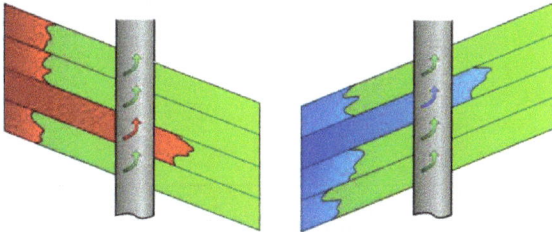

Figure 3.4. Early gas or water breakthrough.

depth is the reason why compact tools are recommended to minimise errors due to this limitation.

3.3.1. *Flow Meters* (*Spinners*)

These are the tools naturally associated with PL. Despite numerous attempts to use other technologies, the spinner-based tools remain the primary way to assess fluid velocities. Even the latest MPT use Micro Spinners (MS) placed at strategic points in the wellbore cross-section.

Spinners are of various types, material and shapes, depending on usage. They turn with as little friction as possible, and include internally located magnets, which will activate powered Hall-effect switches, generating a pulse several times per revolution. If the magnets are somewhat asymmetric, they will give a way for the tool to detect the direction of the rotation.

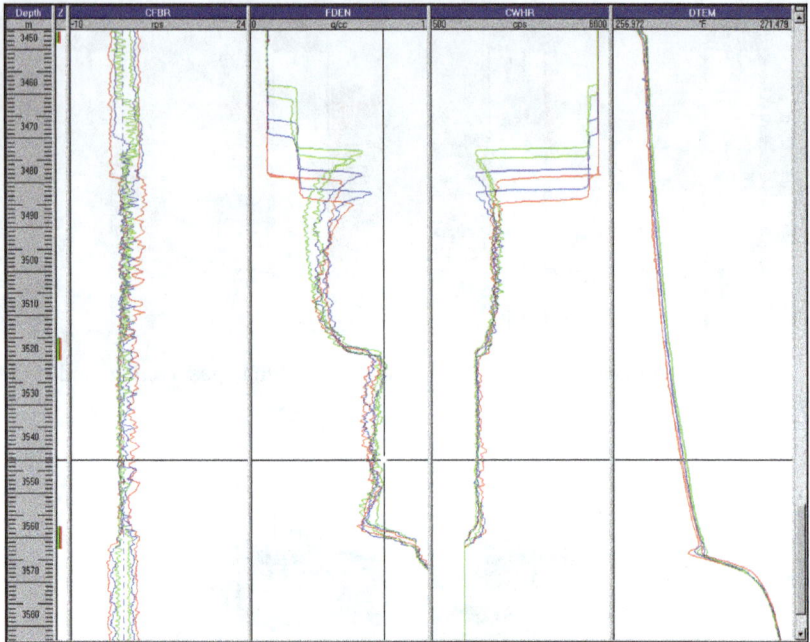

Figure 3.5. Rise of a water column during a shut-in.

The spinners are packaged in several types of tools. There are three main types of flowmeters: inline, fullbore and petal basket (Figure 3.7). Other types are not described here.

Inline flowmeters have small diameters and can be used to log in completions with restricted diameters (tubing, scaled-up wells, etc.). They have a low sensitivity and must be selected to log high-rates/high-velocity wells. Because of the small spinner size, a good centralisation of the tool is required.

Fullbore flowmeters have larger blades that are exposed to a larger part of the flow cross-section. The blades collapse in order to pass the tubing and other restrictions. They expand and start turning when the cross-section becomes large enough. Fullbore flowmeters have a good sensitivity and can be run for a wide range of flow rates and velocities. There may sometimes be issues with injectors, where the blades may collapse when the flow coming from above becomes too large. A lot of tools combine the fullbore spinner with an X–Y

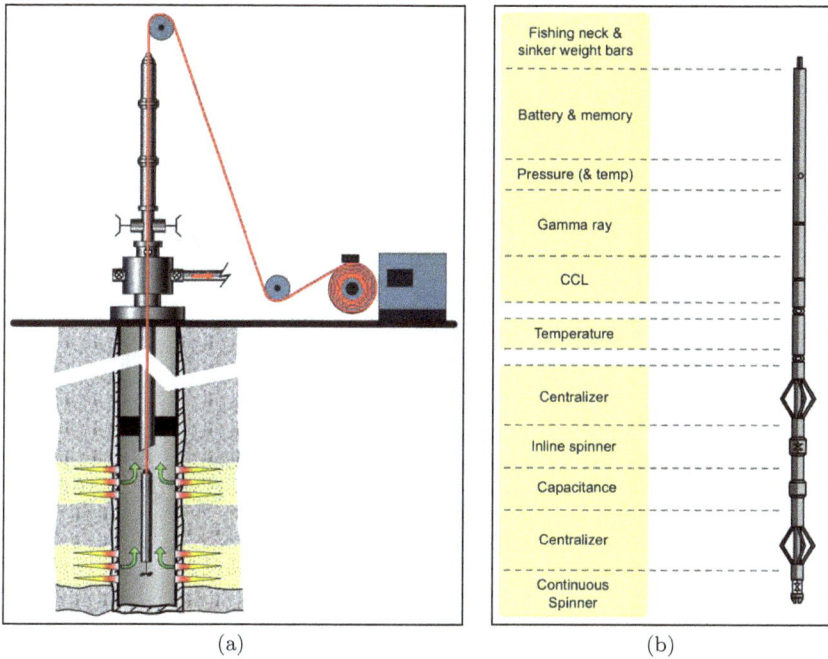

Figure 3.6. Schematics of PL operations (a) and a PL tool string (b).

caliper that will protect the blade, and expand/collapse the tool. Such a setup, combining two tools in one, creates a more compact tool string.

Petal basket flowmeters concentrate the flow towards a relatively small spinner. They are very efficient at low flowrates; however, they are not rugged enough to withstand logging passes and are really designed for stationary measurements, and the tool shape often affects the flow regime.

It is important to realise that spinner-based flowmeters do not measure rates; they do not even calculate the fluid velocity. The output of a spinner-based flowmeter is a spinner rotation in Rotations Per Second (RPS) (or Counts Per Second (CPS) for some tools). The process of converting RPS to apparent velocity, then average velocity, and then ultimately rates, is the essence of PL interpretation and requires additional measurements and assumptions. This is described later in the chapter.

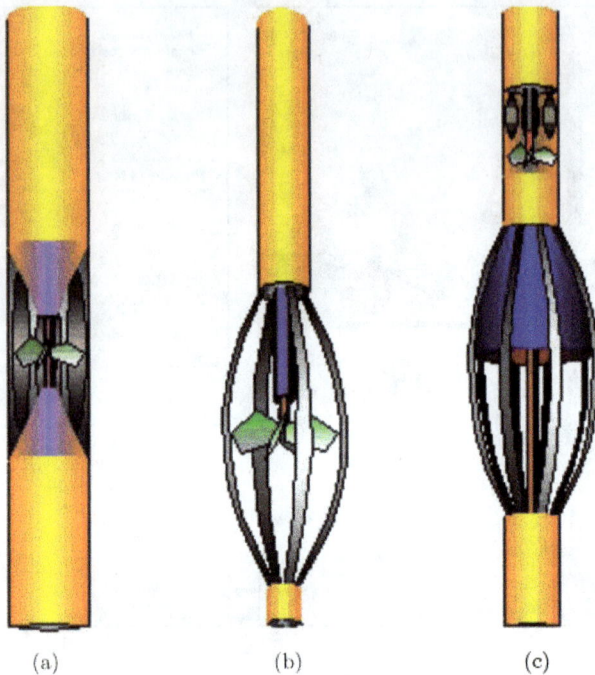

Figure 3.7. Inline (a), fullbore (b) and petal basket (c) flowmeters.

3.3.2. *Density Tools*

In single-phase environments, spinner measurements may get us flowrates. However, when several phases flow at the same time, the problem becomes under-defined, and one needs to get additional measurements in order to discriminate possible solutions.

To schematise, we will need at least one more tool to get two-phase interpretations, and at least a third one to get three-phase interpretations. Without the minimum number of tools, additional assumptions will be needed. If there are more tools than necessary, then one will not be able to match all measurements exactly at the same time, because of the nature of the calculations done (more on this later).

The first natural complement of spinner-type flowmeters are density tools. In a two-phase environment, measuring the fluid density will allow discriminating the light phase and the heavy phase,

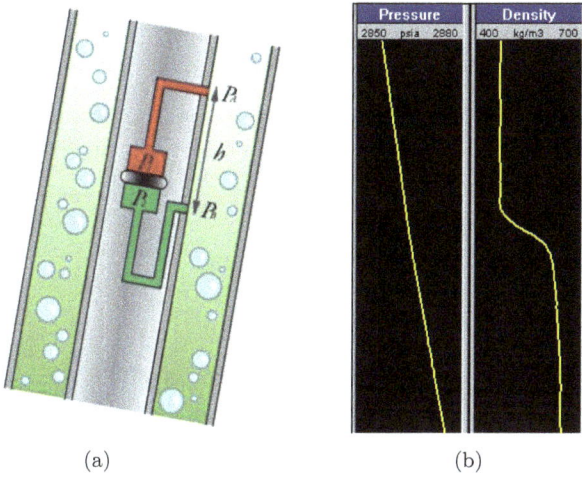

Figure 3.8. Schematics of a gradiomanometer (a) and pressure differentiation (b).

provided that we have a good knowledge of the PVT. There are four main tools that may give a fluid density: gradiomanometers, nuclear density tools, Tuning Fork Density tools (TFD) and pressure gauges after differentiation (Figure 3.8).

3.3.2.1. *Gradiomanometers*

The tools measure the difference of pressure between either sides of a sensing chip. The direct calculation $(P_2 - P_1)$ must be corrected for the hydrostatic pressure of the internal column of silicon oil in order to get the effective pressure $(P_B - P_A)$. This pressure then has to be corrected for deviation and friction effects in order to get a corrected density:

$$\rho_{\text{fluid}} = \frac{[P_2 - P_1] - \Delta p_{\text{fric}} - \Delta p_{\text{acc}}}{gh\cos(\theta)} + \rho_{\text{so}}. \tag{3.1}$$

The acceleration term is generally ignored. For a given surface, the friction gradient is a function of the friction factor (f, calculated from the Reynolds number and the surface roughness), the density, the relative velocity, the friction surface and the flow area:

$$\left[\frac{dP}{dZ}\right]_{\text{friction}} = \frac{f\rho V^2}{8} \times \frac{S}{A}. \tag{3.2}$$

We generally split the friction into tool friction and pipe friction:

$$\left[\frac{dP}{dZ}\right]_{\text{friction}} = \left[\frac{dP}{dZ}\right]_{\text{pipe}} + \left[\frac{dP}{dZ}\right]_{\text{tool}}$$

$$= \frac{f_{\text{p}}\rho V^2}{2} \times \frac{D}{(D^2 - d^2)} + \frac{f_{\text{t}}\rho V_{\text{t}}^2}{2} \times \frac{d}{(D^2 - d^2)}. \quad (3.3)$$

3.3.2.2.   *Nuclear density tool*

This tool sends gamma rays from one side of a chamber and detects them on the other side. The gamma ray attenuation will only be a function of the fluid density inside the chamber. There is no correction for friction or deviation required.

The issue is whether the fluid present in the chamber is representative of the flow through the pipe. The tool very quickly shows its limitations in deviated wells with segregated flow. The existence of a radioactive source is also an issue.

3.3.2.3.   *Tuning fork density*

The TFD tool operates by measuring the effect of the fluid on a resonant fork. As for nuclear density tools, there is no need to correct for frictions and deviation. This is a fairly recent type of tool, and so we will have to wait a little more to assess its efficiency.

3.3.2.4.   *Pseudo-density from pressure*

We calculate the derivative of the pressure with respect to measured depth, and then have to correct for friction and deviation. Generally, this will be used on pressure acquired during slow passes.

3.3.3.   **Capacitance and Holdup Tools**

The holdup of a phase at any given depth is the volume fraction occupied by that phase. Figure 3.9 shows heavy (diagonals) and light (dots) phases and indicates the corresponding holdups.

The holdups are usually labelled Y; they add up to 1 by definition.

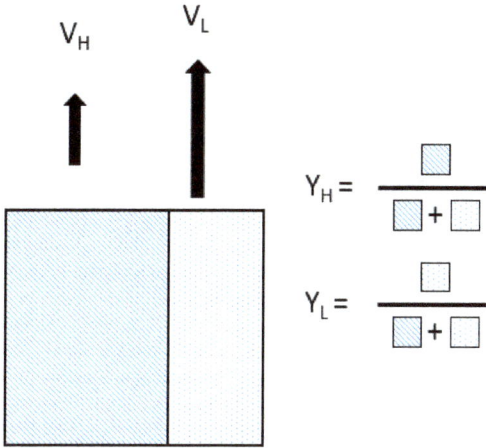

Figure 3.9. Definition of holdup.

Capacitance and holdup tools are designed to provide the holdup of a particular phase. This series of tools is a complement to the spinners in order to differentiate multi-phase flow.

3.3.3.1. *Capacitance tool for water holdup*

This is a tool based on the difference of dielectric constants between water and hydrocarbons. This tool will provide correct measurements when the water holdup is less than 40%. The tool response, given as a calibration curve, is unique and highly nonlinear.

This tool is also subject to delays in the response by filming (down passes) and wetting effects (up passes), hence the risk of the wrong positioning of the fluid contact.

3.3.3.2. *Gas Holdup Tool (GHT)*

This tool is designed to calculate the gas volume fraction in the fluid. A transmitter emits gamma rays; the measurement discriminates the gas based on the amount of backscatter, knowing that the gas has a low electron density and a low backscatter.

The tool gives a measurement across the wellbore with no influence of the formation behind the casing. It is not sensitive to deviation and requires no friction correction.

The negative is that it uses a radioactive source and must be run centralized. Raw counts have to be corrected by the pipe ID, prior knowledge of certain PVT properties, and results may be affected by scale.

3.3.4. *Pressure and Temperature Sensors*

Pressure and temperature measurements are used directly or indirectly, and they constitute two very important components of any PL string.

Pressure is required for PVT calculations; it can be used as an indication of the production stability; it can supplement a missing/faulty density measurement when differentiated; it provides one of the key pieces of information with Selective Inflow Performance (SIP).

Pressure gauges can be split into strain gauges or quartz gauges. With strain gauges, the mechanical distortion caused by the applied pressure is the primary measuring principle. There are several sensor types based on Bourdon tubes, thin-film resistors, sapphire crystals, etc.

In quartz gauges, a quartz sensor oscillates at its resonant frequency. This frequency is directly affected by the applied pressure (Figure 3.10).

Like the pressure, the temperature is used in PVT calculations. It can also reveal flow outside the wellbore, because of cement channelling leak for instance. The temperature can be used quantitatively, provided an adequate forward model for calculations is available.

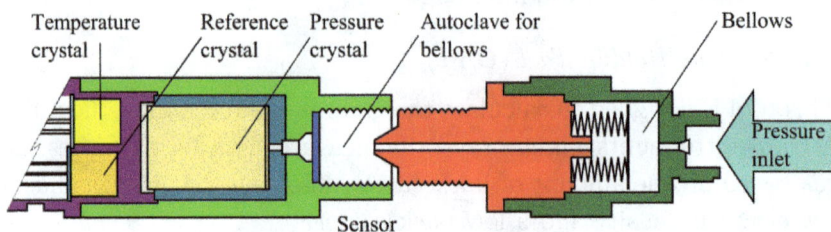

Figure 3.10. Quartz sensor example.

3.3.5. *Depth and ID Devices*

3.3.5.1. *Depth measurement*

The depth is measured at the surface by measuring the length of the cable run in-hole. The depth measurement does not consider possible stretch of the cable or, conversely, slack, due for instance to deviation or restrictions. A tension measurement can be used to spot such cases. As mentioned earlier, the log data need to be offset in time to be displayed at the same depth. This is because the various sensor measurement points are at different depths.

3.3.5.2. *Depth correction: Open-hole gamma ray*

The tool '0' is set at surface when the log is run. The first task in the data Quality Assurance/Quality Check (QA/QC) is to set the log data consistently with the other available information: completion, perforations, etc. This can be achieved by loading a reference open-hole gamma ray and shifting the acquired data so that the PL and open-hole curves overlay. The signal may not be strictly the same due to completion, scale, etc.

3.3.5.3. *Depth correction: Cased-hole CCL*

An alternative to the gamma ray for depth correlation is a Casing Collar Locator (CCL), a measurement that will react in front of the casing collars at known depths.

3.3.5.4. *ID calculation: Calipers*

Calipers are mechanical devices used to calculate the cross-section of the wellbore. They are critical since the cross-section must be known to convert velocities to flowrates. Even in cased holes, a completion diagram may not reflect the reality. Calipers can be integrated in the spinner tool or as a separate device. They usually measure the diameter in two orthogonal directions; in this case they are referred to as X–Y calipers. For such calipers, the ID at every depth is calculated as $\sqrt{(X^2 + Y^2)}$.

3.4.   A Typical PL Job Sequence

The basic assumption in PL is that the well is in steady state. It is therefore important that the well be stabilized before running the tools. A typical job will consist of several surveys corresponding to different surface conditions. Shut-in surveys are also recorded, with the goal being to calibrate the tools in an environment where the phases are segregated. Shut-ins can also reveal crossflow due to differential depletions, provide a reference gradient in shut-ins and provide a baseline for a number of measurements in flowing conditions, e.g. temperature.

Part of the job planning is to account for the time it will take for the well to be stable. The notion of stability is defined as a pressure variation over time since we know from Welltest that the flowing pressure will usually not be strictly constant. Time lapsed passes will be interesting to record evolution with time, for instance, the warmback following an injection period as shown in Figure 3.11.

3.4.1.   *Multirate PL and SIP*

In order to get SIP, it is necessary to achieve more than one rate. Typically, three rates and a shut-in are recorded, as illustrated in Figure 3.12.

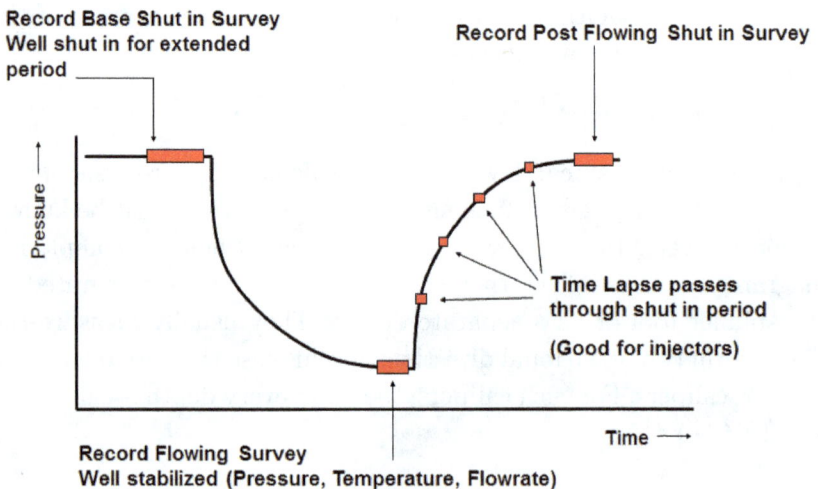

Figure 3.11. A typical PL job sequence.

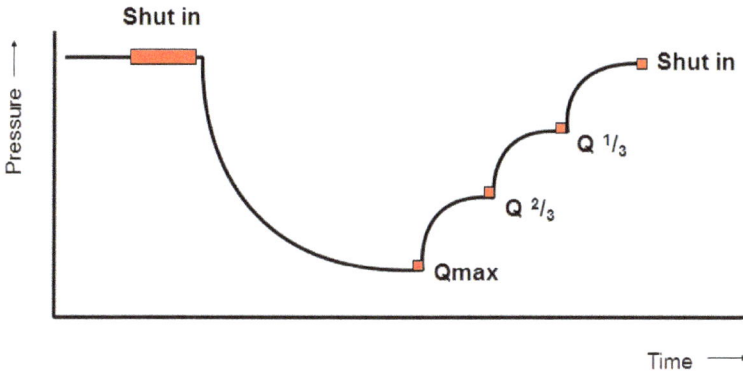

Figure 3.12. Multirate PL.

3.5. Typical Job

In a single-phase situation, a typical tool string will comprise temperature and pressure, spinner, and caliper. No further information is required, assuming of course that the flow conditions are indeed single-phase. It is not uncommon to encounter fluids downhole that are not produced at the surface. If in doubt, it is always better to add a fluid identification tool in the string (density or holdup) before.

In a multi-phase situation, there are $n-1$ additional unknowns, n being the number of phases. So in two-phase flow a density or holdup is required, and in three-phase two such independent measurements are required.

For all tools but the spinner, one pass would be sufficient for calculation. However, comparing several passes for other tools is a way of judging the well stability. Having multiple passes also provides more chances to have a representative measurement if the data are bad on some section of some passes.

The spinner calibration, which is explained next, requires several passes at various logging speeds. A typical job will comprise three or four down and three or four up passes, as illustrated in Figure 3.13. Passes are normally numbered by increasing speed, and the slow passes are recorded first. This means that Down 1 is normally the first and slowest pass in the well.

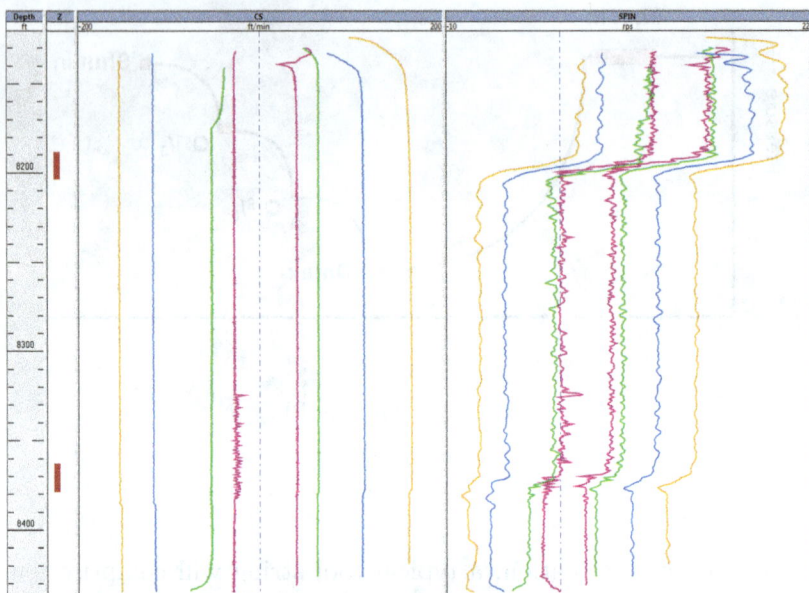

Figure 3.13. Spinner and Cable Speed (CS) for an eight-pass job.

Stations may be recorded for certain tools that require it. In addition, the ability to display the measurement of a station versus time is a further indication of well stability or instability.

3.6. Data Loading, QA/QC, and Reference Channels

The first task after the job is run is to load and review the log data for QA/QC. The following is a list of checks and possible actions.

Data editing — General: Telemetry errors can introduce spikes on data that in turn will upset the scaling and make the QA/QC difficult (Figure 3.14). De-spiking should be achieved beforehand to avoid this problem, using for instance a median filter. Some tool responses may also be noisy, for instance nuclear tool responses, and will usually deserve a processing with a low-pass filter.

Data editing — Spinner/CS: When starting fast passes, "yo-yo" effects may be seen. Similarly, oscillations may appear on the flowmeter as a result of the cable sticking/slipping along the length

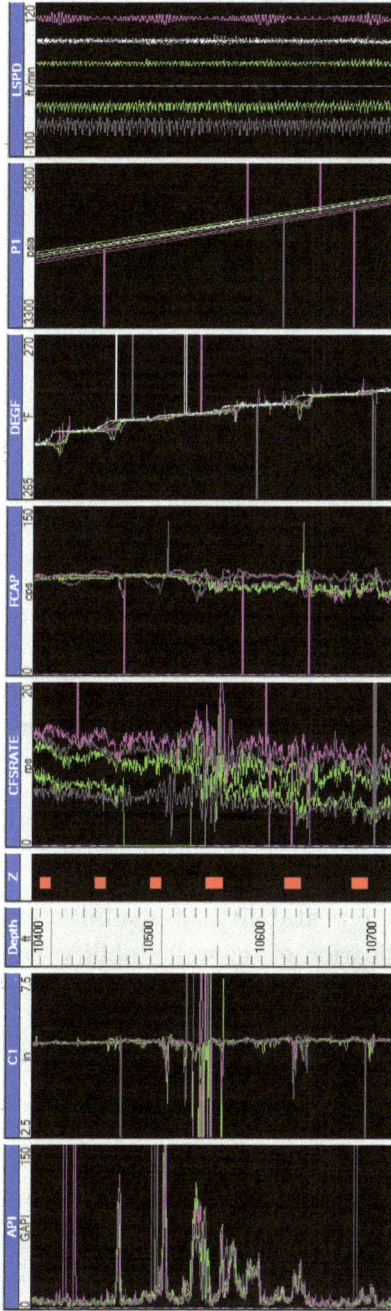

Figure 3.14. Example of telemetry spikes.

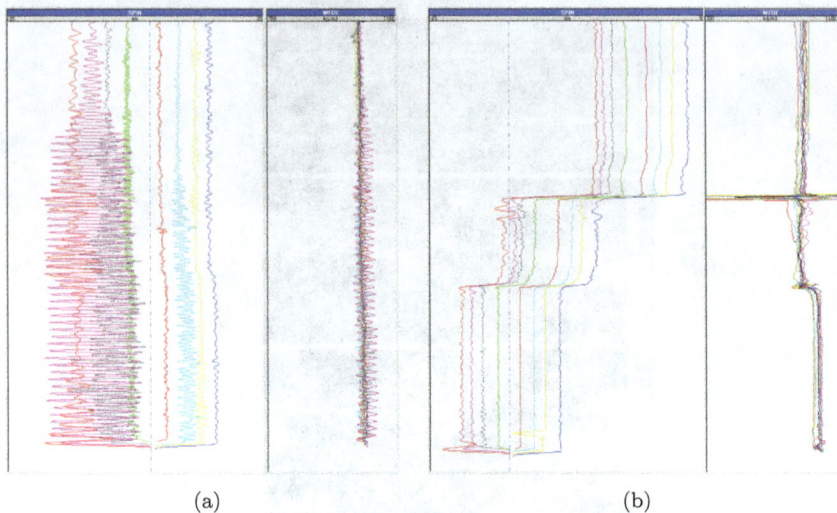

(a)   (b)

Figure 3.15. Stick and slip (a) and yo-yo at start of passes (b).

of the completion. Those can be edited out with a sliding window averaging (Figure 3.15).

Unsigned spinners will need to be corrected before processing, on up passes in producers and down passes on injectors (Figure 3.16).

Depth matching: Using the gamma ray or CLL, all the available data should be set with coherent depth. Sometimes the depth correction might require more than a shift.

Repeatability: The repeatability of readings from one pass to another will be an indication of the well stability. Some tool readings are affected by the CS when the fluid mixture changes drastically (density or temperature for instance), and this should be noted as it will orient the choice of reference channels for the calculations.

Consistency: A first consistency check between sensors can be done at this stage. Beware that some tools will need corrections before they can provide a quantitative answer (e.g. a pseudo density — dP/dZ — in a deviated well). Some tools react to what happens behind the tubing or casing (e.g. the temperature) and will behave differently from the spinner.

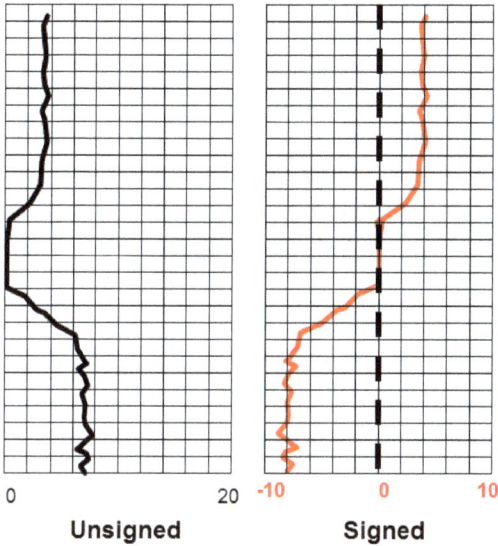

Figure 3.16. Comparison of unsigned/signed responses.

Qualitative analysis: Once all data have been cleaned, most of the questions can be answered qualitatively, but there are some pitfalls, like mistaking a change in spinner response to an inflow/outflow when it is due to a change of ID.

Reference channels: For any measurement that will be used quantitatively in the interpretation, a single curve must be selected or built. Most of the time this curve will be the slowest down pass, but some editing or averaging may be necessary. This procedure will not apply to the spinner that needs to be calibrated first, in order to calculate velocities, before we can define or build the reference velocity channel.

3.7. Spinner Calibration and Apparent Velocity Calculation

To conduct a quantitative interpretation, the spinner output in RPS needs to be converted into velocity. The relation between RPS and velocity depends on, amongst other things, the fluid property, and for this reason, an *in situ* calibration is required.

3.7.1. *Spinner Response*

The spinner rotation depends on the fluid velocity relative to the spinner; this is a function of the fluid velocity and the tool velocity. The usual sign conventions consider that the tool velocity is positive, while going down and negative when coming up. Similarly, the spinner rotation is counted positive when the fluid is seen by the spinner as coming from below, and negative when it is seen as coming from above. With these conventions, the spinner rotation is relative to the sum CS + fluid velocity.

The response of an ideal spinner run in a static fluid would be as plotted in Figure 3.17, with two distinct response lines for up passes (negative CS) and down passes (positive CS).

1. The RPS value is a linear function of the velocity; the response slope depends on the spinner pitch, i.e. its geometry only.
2. The spinner rotates with the slightest movement of the tool, i.e. the slightest movement of fluid relative to the tool.
3. The negative slope is lower, typically as the tool body acts as a shield; this is not the case with a symmetrical tool like an in-line spinner.

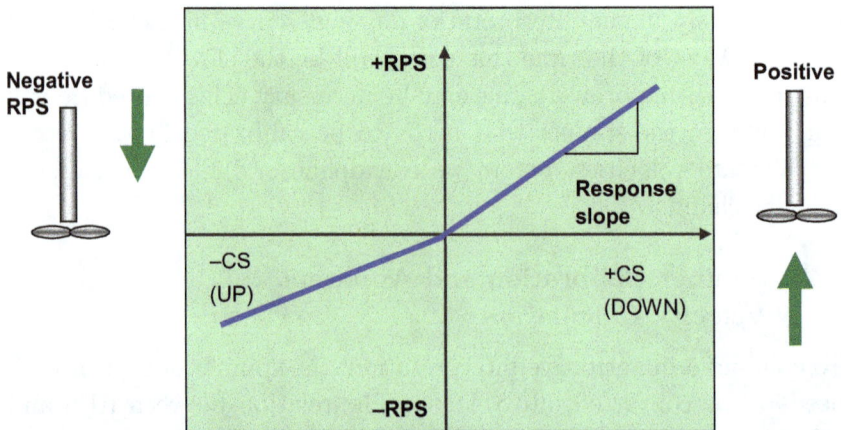

Figure 3.17. Ideal spinner response (in a no-flow zone).

In reality, the response is affected by the fluid properties as well as the bearing friction. Equation (3.4) is a possible representation [1]:

$$\text{RPS} = aV_{\text{fs}} - \frac{b}{\rho V_{\text{fs}}} - c\sqrt{\frac{\mu}{\rho V_{\text{fs}}}}. \tag{3.4}$$

For PL interpretation, we will consider that the calibration is still a straight line. Since this line is approximating a non-linear function, it may vary with the encountered fluid. In addition, the tool response is shifted by a threshold velocity, the minimum velocity required for the spinner to rotate. This threshold velocity will depend on the fluid; typical numbers for a fullbore spinner are 3–6 ft/min in oil and 10–20 ft/min for gas.

Figure 3.18 represents the spinner response in a no-flow zone as a function of the CS. If the fluid is moving at some velocity, V_{fluid}, then the tool response will be the same, but shifted to the left by V_{fluid} as shown in Figure 3.19. The reason behind the shift is that since the spinner reacts to the sum of $(V_{\text{fluid}} + \text{CS})$, the RPS value for a CS value "X" in V_{fluid} is the response to a CS value of $(X + V_{\text{fluid}})$ in the no-flow zone.

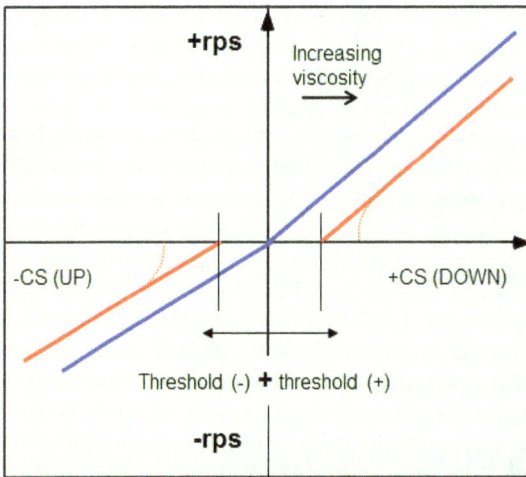

Figure 3.18. Real spinner response (in a no-flow zone).

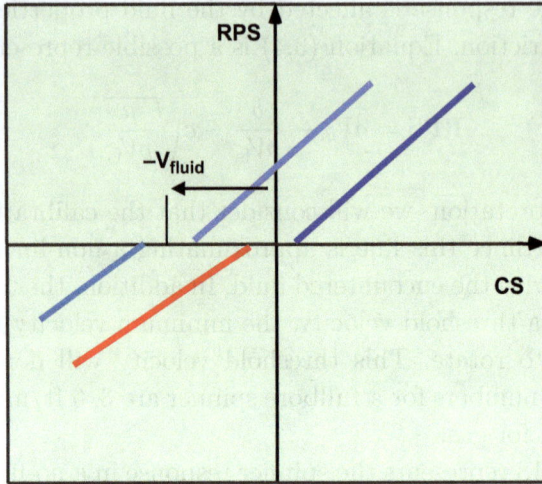

Figure 3.19. Real spinner response in a no-flow zone, and a zone with production.

3.7.2. *Spinner In Situ Calibration*

In practice, the objective is to build the calibration response *in situ* to take into account the changing fluid properties and their effect on the spinner calibration. In Figure 3.20, the fluid velocity is represented on the left. Six passes have been run, and the spinner response is shown on the right.

Three stable intervals are represented by the sections in blue, green and red. The corresponding points are plotted on an RPS versus nonlinear cable speed plot (Figure 3.21) and the lines are drawn by linear regression.

Historically when doing hand calculations, the usual method was to consider a spinner calibration zone in stable regions in between every perforation, as shown earlier. The velocity was then calculated directly from the cross-plot for every zone. Today's idea is that you only put in spinner calibration zones because you think there has been a change of slope or threshold (usually due to a change of fluid type). In theory, a single-phase well only needs one spinner

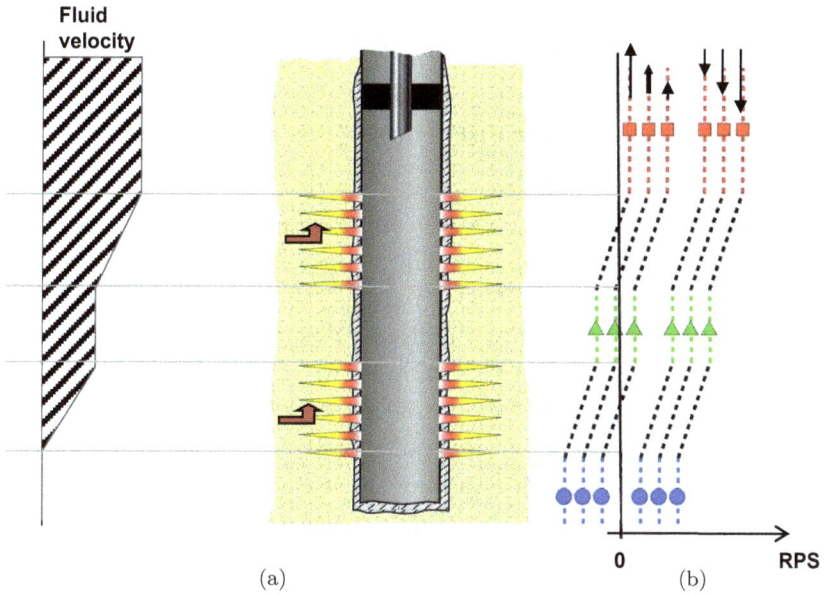

Figure 3.20. Fluid velocity (a) and spinner response (b) for the six passes.

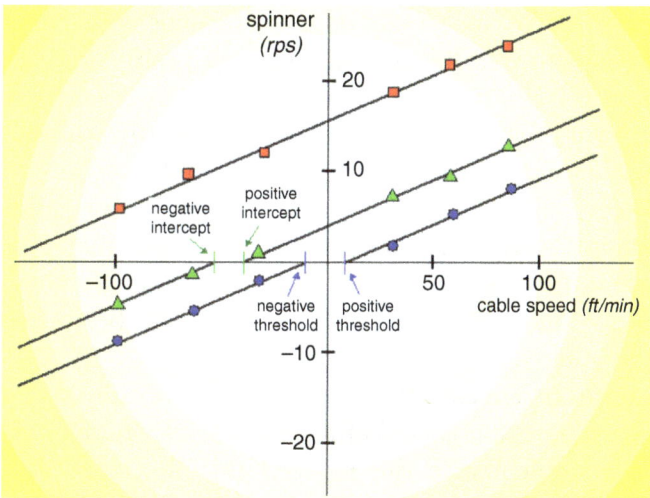

Figure 3.21. *In situ* calibration.

calibration zone. In reality, having multiple zones — as long as they are stable — ensures that any change is properly captured.

3.7.3.   *Velocity from Calibration: Threshold Handling and $V_{apparent}$*

The calibration does not give the fluid velocity directly, and some calculation or assumptions remain to be done. If the response of the various sections were strictly parallel, then we would know from the previous discussion that the fluid velocity could be obtained by estimating the horizontal translation between, say, the positive line of a given zone and the positive line of the no-flow zone. This method is fine for manual analysis, but it is quite limitative. A general approach needs a systematic way of handling slope and threshold variations.

3.7.3.1.   *Threshold options*

Below are options usually considered.

1. **Unique value of $(+)$ and $(-)$ thresholds for all zones:** The apparent velocity for a point on a positive line is calculated based on the slope of that line and the common positive threshold. The apparent velocity for a point on a negative line is calculated based on the slope of that line and the common negative threshold. This mode is suitable in case of single-phase fluid.
2. **Distinct thresholds, unique ratio threshold $(-)/$ [Intercept $(-)$ − Intercept $(+)$]:** This ratio is equal by default to $7/12 = 0.583$, but can be set from the value of a no-flow zone. Obviously, this can be used only on zones with both positive and negative intercepts.
3. **Independent thresholds:** This mode allows different thresholds for each calibration zone and is the most general one. Note that the only problem with this mode is that in a zone where there is fluid movement, at best we can get the sum of the thresholds. Deciding the positive and the negative can be done bluntly (i.e. halving the sum) or based on the split on the no-flow zone.

3.7.3.2. *Apparent velocity*

No matter what method is used, we will end up with a measurement of velocity that represents the fluid velocity seen by the spinner. This value is the average velocity in the cross-section covered by the spinner and is different from the actual fluid average velocity. For this reason, it is referred to as apparent velocity, noted V_{apparent} or V_{APP}.

3.7.4. *Apparent Velocity Log*

After the calibration is complete and the threshold mode has been selected, the goal is to obtain a continuous apparent velocity curve representing the velocity at every depth.

The slope and threshold values for a calibration zone will apply to all calculations performed within that zone. Data above the top calibration zone will use the top calibration, and data below the bottom calibration zone will use the bottom calibration. In between consecutive calibration zones, the possible options are to move linearly from the values of a zone to the value of the other or to make this change occur sharply. From the previous discussion on spinner response we know that the difference in spinner response is caused by the change in fluid. Because the change in fluid property is local (to a particular in-flow zone) rather than spread over a large interval, a sharp transition is probably more liable. All the possibilities should be offered: progressive change, sharp change or a mix of both.

Having decided how to obtain a slope and threshold for any depth, the procedure to get the apparent velocity curve for a given pass is as follows: for a given measuring point, we have the CS and the spinner rotation; from graphing the point on the calibration plot, we follow the slope, get the intercept and correct by the threshold (Figure 3.22).

Producing one apparent velocity channel per pass allows an overlay and is a further check of the well stability and the calibration adequacy. When this check has been made, the multiple apparent velocity curves are typically replaced with a single average value (median stack or lateral average). This single apparent velocity curve is the sole information required for quantitative analysis.

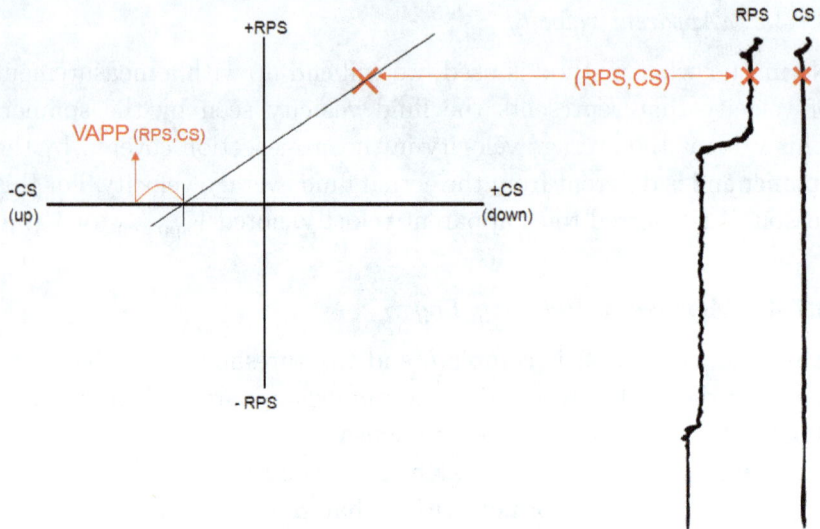

Figure 3.22. Calculation of apparent velocity.

3.8.   Single-Phase Interpretation

Rate calculations may be performed on each depth frame or averaged on calculation zones of interest. Such calculation zones could be the calibration zones, the top of each perforation zones, but in most cases the engineer will define where the rate calculations are relevant.

The spinner calibration allows us to get the apparent velocity, V_{APP}, everywhere there is a measurement (Figure 3.23).

To get a single-phase rate we need the total average flow velocity, which can be expressed from V_{APP} with some correction factor, usually noted as Velocity Profile Correction Factor (VPCF):

$$V_M = \text{VPCF} \times V_{APP}. \qquad (3.5)$$

3.8.1.   *Velocity Profile Correction Factor*

Historically, and at least for any manual interpretation, the VPCF is taken as 0.83. More generally, this factor can be calculated from the Reynolds number and the ratio of blade diameter to pipe diameter using the correlation illustrated in Figure 3.24.

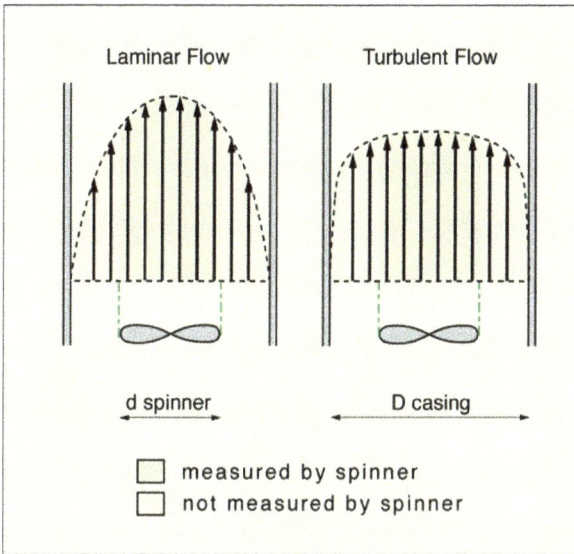

Figure 3.23. Velocity profiles and spinner sampling section.

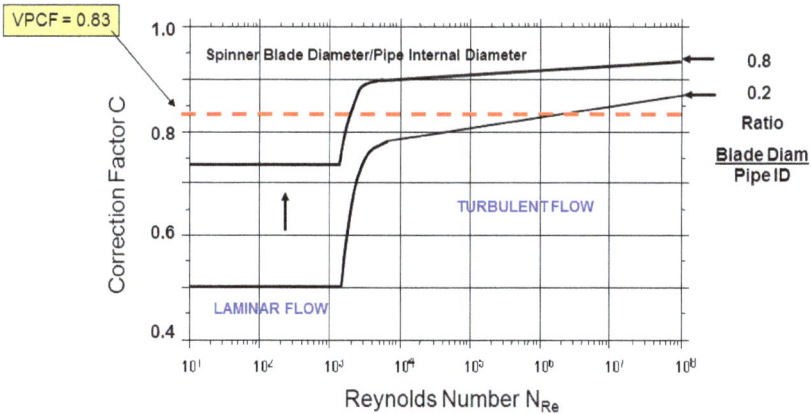

Figure 3.24. VPCF versus Reynolds number for different ID ratios.

Under a value of 2,000 for the Reynolds number, the flow is laminar with a parabolic velocity profile. In this situation, the maximum velocity is twice the average, leading to VPCF = 0.5 for a small blade diameter. When the Reynolds number increases, the correction factor increases from 0.5 and its value tends asymptotically to 1. Also, as the blade diameter tends to the pipe ID, the correction factors moves towards 1.

The Reynolds number, N_{Re}, is expressed below for fluid density, ρ, in g/cc, diameter, D, in inches, velocity, ν, in ft/sec and viscosity, μ, in cp:

$$N_{\mathrm{Re}} = 7.742 \times 10^3 \frac{\rho D \nu}{\mu}. \tag{3.6}$$

Obviously, the value we are seeking — the fluid velocity — is part of the equation, meaning that an iterative solution is required.

The classical solution is to assume a value of velocity, typically based on a VPCF of 0.83, then calculate the Reynolds number, from which a new value of VPCF would be calculated, hence a corrected estimation of the flow velocity.

The process would go on until the solution eventually converges (Figure 3.25).

In modern software, this has been replaced by a regression algorithm, and the single-phase calculation is only a specific case of what is done for much more complicated processes like multi-phase rate calculations or the processing of MPT.

The principle of the non-linear regression process is that we take as unknowns the results we wish to get, here the single-phase downhole rate, Q.

The target will generally be the observed tool measurement.

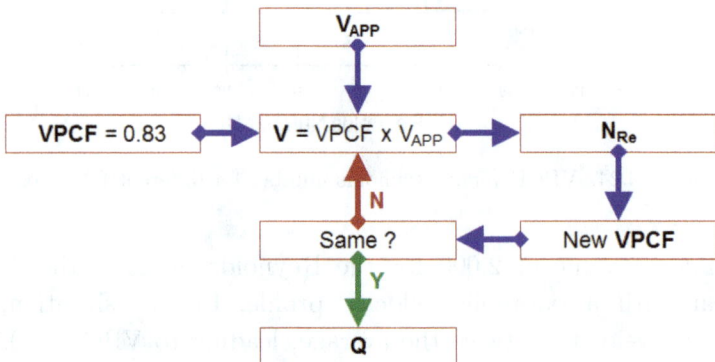

Figure 3.25. Single-phase interpretation workflow.

In the case of a single-phase calculation, the target is the apparent velocity calculated after the spinner calibration.

From any value of Q in the regression process, we calculate the velocity, hence the Reynolds number, hence the VPCF, hence a simulated apparent velocity.

This allows the creation of a function, $V_{APP} = f(Q)$. We then solve for Q by minimizing the standard deviation between the simulated apparent velocity and the measured apparent velocity:

$$Q \rightarrow v \rightarrow N_{Re} \rightarrow \text{VPCF} \rightarrow V_{APP}$$

Simulated Apparent Velocity: $V_{APP} = f(Q)$

Measured Apparent Velocity: V_{APP}^*

Minimize Error Function: $\text{Err} = (V_{APP} - V_{APP}^*)^2$. $\qquad$ (3.7)

3.8.2. *Single-Phase Interpretation Results*

Figure 3.26 is a typical presentation of results.

In this water injector, five different calculation zones (grey) were selected to isolate the contributions. The non-linear regression described previously was executed on all of them to get Q. The results are summarised by the QZT (Z: Zoned, T: Total) track where the value of a given calculation zone is extended up to the in-flow zone above, and down to the in-flow below.

In-flow zones (black) can be distinct from the perforations (red), simply to capture the fact that not all the perforation intervals may produce, or in this case take, fluid.

The QZI (Z: Zoned, I: Incremental) track represents contributions or injections, and they are obtained by the difference of the rate above and below an inflow.

The last rate track, noted Q, represents the application of the non-linear regression at every depth, the point being to obtain a continuous log everywhere faithful to the log data.

Having this log provides in particular a guide to refine the position of calculation zones.

In the complete log (Q) or for the zone rates (QZT), the calculation of the rate at depth 1 is independent of the calculation

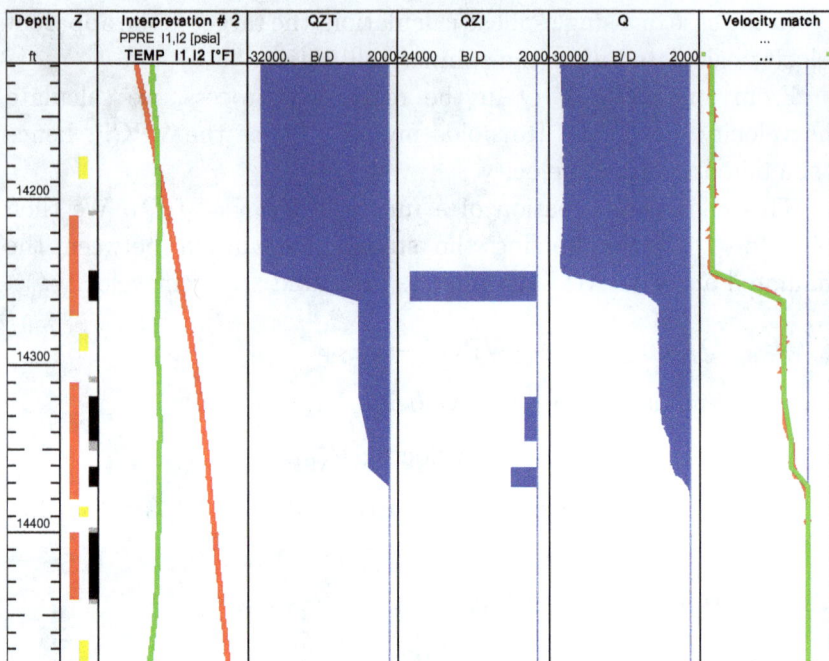

Figure 3.26. Typical log presentation for a single-phase interpretation.

at depth 2. As a result, those rates may entail contributions of a sign or amplitude that is not physically justified. We can address this potential inconsistency in a global regression process described later in this chapter.

The final track above shows the target V_{APP} and the simulated equivalent in green. Note that we arbitrarily decided to take the apparent velocity as the target function, rather than the real tool response in RPS.

We could have integrated the spinner calibration in the regression process and matched the RPS measurements for the different selected passes.

The simulated (green) curves and the QZT logs are referred to as "schematics".

3.8.3. *Matching Surface Conditions*

It is possible to apply a global gain on the rate calculations.

This may be relevant if one wants to match the production above the top producing zone with the measured surface rates (if one relies on this). Numerically, it amounts to allowing a multiplier to the VPCF.

3.9. Multi-Phase Interpretation

In single-phase interpretation, the spinner alone is providing the answer, even if the determination of the correction factor needs a good grasp of the downhole conditions to have a representative density and viscosity. In multi-phase flow, there are at least as many unknowns as we have phases, i.e. one rate per phase. Actually, due to the fact that different phases do not flow at the same velocities, the number of unknowns is larger, including not only rates but also holdups.

3.9.1. *Definitions*

3.9.1.1. *Holdups*

This definition was given before, but it is repeated for clarity. The holdup of a phase is the volume fraction occupied by that phase. Figure 3.27 shows heavy (diagonals) and light (dots) phases and indicates the corresponding holdups.

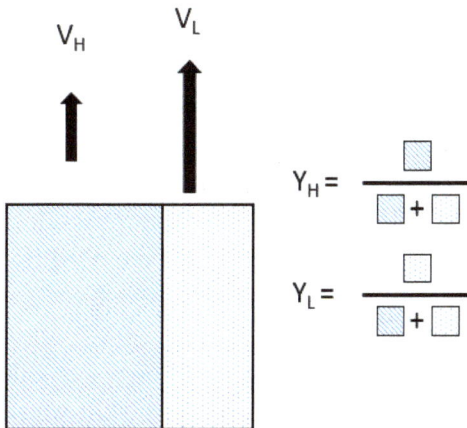

Figure 3.27. Definition of holdup.

The holdups are usually labelled Y; they add up to 1 by definition.

In two phases with a heavy (H) and a light (L) phase,

$$Y_H + Y_L = 1. \tag{3.8}$$

In three phases with water (w), oil (o) and gas (g),

$$Y_w + Y_o + Y_g = 1. \tag{3.9}$$

3.9.1.2. *Phase velocities*

The average velocity of a particular phase is obtained from the rate of that phase, its holdup, and the cross-sectional area by

$$V_p = \frac{Q_p}{A \times Y_p}. \tag{3.10}$$

3.9.1.3. *Slippage velocity*

The slippage velocity is the difference between the velocities of two distinct phases. When light and heavy phases are considered, the slippage velocity is usually defined as the difference between the light phase velocity and the heavy phase velocity, namely

$$V_s = V_L - V_H. \tag{3.11}$$

When going uphill, the light phase will move faster and V_s will be positive. The opposite situation will be encountered when going downhill, where the heavy phase will go faster. The slippage velocity is not something the PL tool measures, nor are the phase rates (at least with conventional tools). Getting the rate values will only be possible if we can estimate the slippage value using a correlation. There are many correlations available in the literature, empirical or more rigorously based. For the time being, let us simply assume that such correlation will be available to us if we need it.

3.9.2. *Starting with Two-Phase*

For two-phase flow, the alternatives will be water–oil, water–gas and oil–gas. Even though the general approach that we advocate is using non-linear regression, we describe here a deterministic approach, the value of which is to explain the concepts in a simple situation and to introduce the basic notions/presentations that will be used in the general case. Recalling the definitions mentioned above and using the subscript H for heavy and L for light, we can write:

$$Y_H + Y_L = 1, \tag{3.12}$$

$$Q_H + Q_L = Q_T, \tag{3.13}$$

$$V_S = V_L - V_H = \frac{Q_L}{A \times Y_L} - \frac{Q_H}{A \times Y_H}$$

$$= \frac{Q_T - Q_H}{A \times (1 - Y_H)} - \frac{Q_H}{A \times Y_H}. \tag{3.14}$$

Solving for Q_H gives

$$Q_H = Y_H \times [Q_T - (1 - Y_H) \times V_S \times A] \tag{3.15}$$

and to finish,

$$Q_L = Q_T - Q_H. \tag{3.16}$$

Holdup is a quantity we can measure directly or infer from density. If we measure mixture density and know the individual phase densities downhole, then holdup can be obtained as shown in Eq. (3.17):

$$\rho = \rho_H Y_H + \rho_L Y_L \Rightarrow Y_H = \frac{\rho - \rho_L}{\rho_H - \rho_L}. \tag{3.17}$$

Note that when density is measured with a gradio, the reading needs to be corrected for frictions. This correction requires the knowledge of velocity and the fluid properties and so, like in single-phase, such a calculation will necessitate an iterative solution scheme.

Since we know how to calculate the total rate from the single-phase equivalent, all we need is a way of determining the slippage velocity. In the simplest situations, this could be done manually. Figure 3.28 shows the Choquette correlation, representative of bubble flow in a vertical well for an oil–water mixture.

With such a graph, the slippage is obtained from the holdup and density difference. With a spinner and a density measurement, the

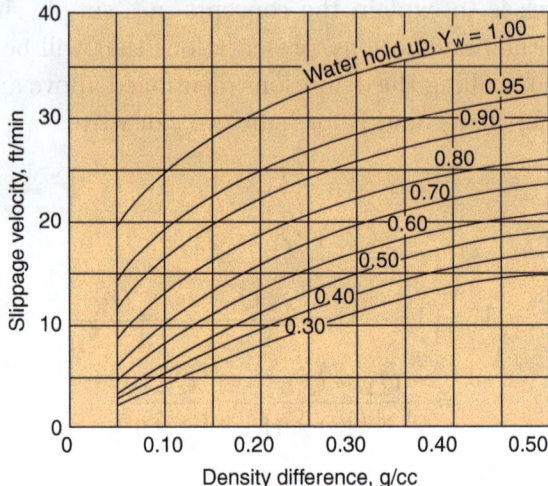

Figure 3.28. Choquette bubble flow graph.

steps of a manual deterministic approach would be straightforward:

1. estimate Q_T from V_{APP} and VPCF = 0.83;
2. estimate Y_H from ρ and ρ_H and ρ_L. Iterate for frictions and VPCF as desired;
3. get V_s from the Choquette chart;
4. estimate Q_H and thus Q_L.

In a general situation, the slippage velocity is given as a function of the rates and the fluid properties (densities, viscosity, surface tensions, etc.). The manual approach must be modified.

3.9.3. *General Solution Using Non-Linear Regression*

The suggested workflow, as in single-phase, relies on using non-linear regression to find the rates that minimise an objective function defined by the error between the target measurements and the simulated equivalent (Figure 3.29). In the case above, for instance, all we need is a forward model that from an assumption of the rates Q_H and Q_L calculates the simulated V_{APP} and density. In a general

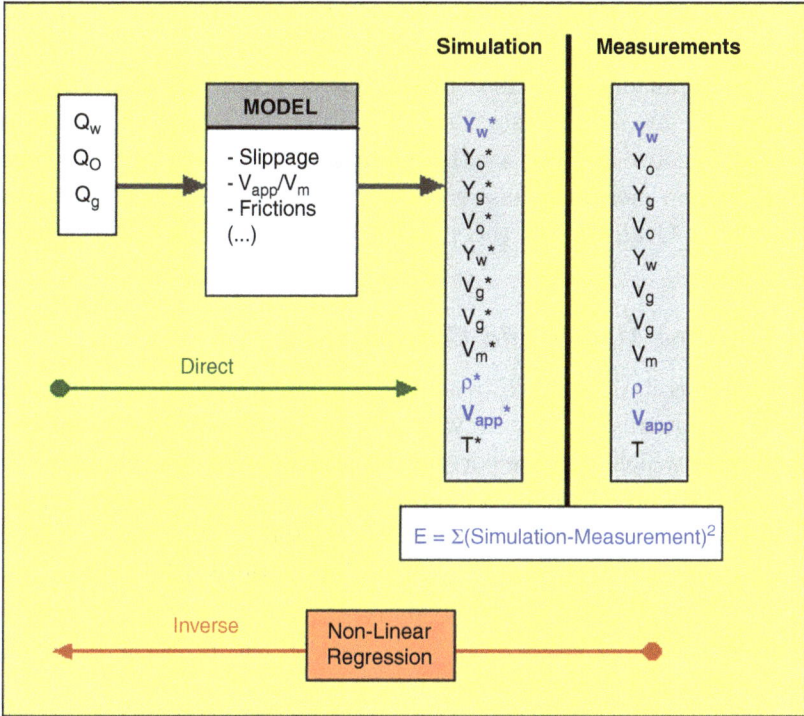

Figure 3.29. Methodology using non-linear regression.

situation, the steps are as follows:

1. from the rates, the fluid and the local geometry, get the slip velocities;
2. from the slip velocities and the rates, get the holdups;
3. from the holdups, calculate the fluid mixture properties;
4. calculate the simulated tool response using the relevant model (e.g. VPCF, friction equations, etc.).

The interest of such an approach is that any number and type of measurements can be chosen as targets provided that they are sufficient and the problem is not undetermined (insufficient measurements for the number of unknowns). Having redundant measurement is also possible, where the regression will try and find a compromise based on the confidence assigned by the user to the different measurements.

It is important to realise in the above procedure that slippage correlations are a necessary evil in order to relate the rates to holdups, because holdups are the quantities we are best at measuring. When PL can measure phase rates or phase velocities directly, which is the Holy Grail of PL, the procedure can do away with slippage correlations.

3.9.4. *Flow Models and Correlations*

Slippage correlations have been derived for a number of situations either empirically, or by solving the general momentum balance equation in which case the correlation is said to be "mechanistic". In general, the behaviour of the slippage velocity will strongly depend on the nature of the flow regime encountered. Whereas a water–oil mixture moving upward can often be considered in bubble flow, liquid–gas mixtures will give rise to much more complex regimes, as illustrated in Figure 3.30 with one of the classical empirical correlations by Duns and Ross.

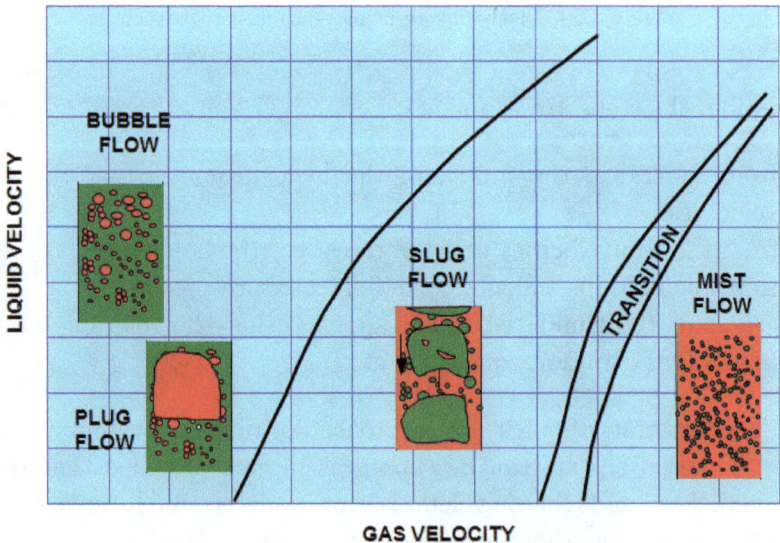

Figure 3.30. Duns and Ross flow map.

In each regime, a specific slippage correlation will be applicable with extreme cases like mist flow, where no slip exists between the two phases. Many correlations will start with a determination of the flow regime using a flow map and continue with the application of the corresponding slippage equation. Slippage correlations can be organised into the following categories:

Liquid–liquid: This category gathers bubble flow correlations (e.g. Choquette), correlations developed for stratified flow (e.g. Brauner) and combinations of those.

Liquid–gas: This is by far the largest population, with many empirical correlations (Duns and Ross, Hagedorn and Brown, Orkiszewksi, etc.) and mechanistic models (Dukler, Kaya, Hassan and Kabir, Petalas and Aziz, etc.). Most of those correlations have been primarily designed for representing wellbore pressure drops in the context of well performance analysis.

Three-phase: There are very few such correlations, and when they exist they are for a very specific situation, e.g. stratified three-phase flow (Zhang). In practice, three-phase slippage/holdup prediction from rates are done using two two-phase models, typically for the gas–liquid mixture on one hand and the water–oil mixture on the other.

3.9.5.   *Graphical Presentation*

A graphical presentation, albeit based on a two-phase analogy, can help understand some previous concepts and provides a means of comparing different correlations.

For a given mixture rate Q, we consider the possible mixture and plot the corresponding mixture density. The x-axis values range from 0 (100% light phase) to Q (100% heavy phase). The density at those end points is the relevant single-phase density. The equation derived previously for Q_H is now used to express Y_H:

$$Y_H = \frac{Q_H}{[Q_T - (1 - Y_H) \times V_S \times A]}. \tag{3.18}$$

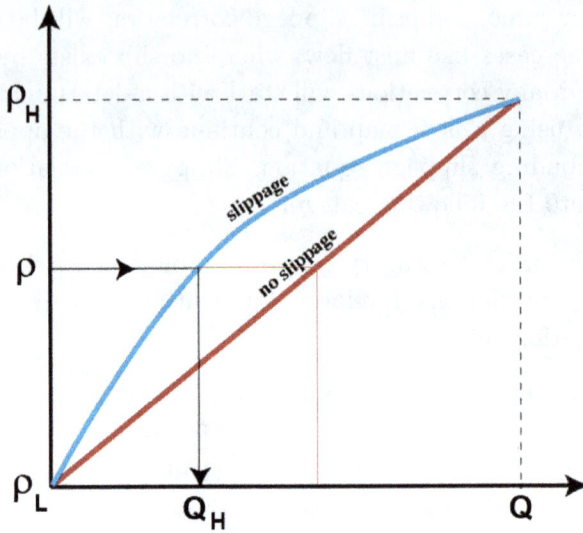

Figure 3.31. Density versus heavy phase rate with and without slippage.

Equation (3.18) shows that, without slippage, the heavy phase holdup would be equal to the heavy phase cut. This situation is represented by the red line in Figure 3.31. With slippage velocity on the other hand, V_s is positive uphill; hence the equation above tells us that Y_H should be higher than the cut. The curve representing the case with slippage is thus above the no-slip line. An opposite situation would be expected downhill.

Another way of considering the plot in Figure 3.31 is to realise that for a given cut, the higher the slippage, the heavier the mixture will be. This is because the light phase is going faster and therefore occupies less volume in the pipe. The light phase is "holding up" the heavy phase, leading to more heavy phase present at any depth than there would be with no slippage. For a given solution with slippage, the density will read heavier than if there is no slip.

We can look at this plot from a final perspective. If we have a measurement of density, the solution we should find (or that the nonlinear regression will find) is seen graphically by interpolating the relevant curve for this density value.

3.9.6. *Zoned Approach and the Zone Rates Plot*

The simplest regression-based interpretation method, called the zoned approach, amounts to using the regression described previously on a set of user-defined calculation zones. As in single-phase, the zones are selected in the stable intervals above and below the in-flow zones.

For each zone, a non-linear regression is performed, and the result of this regression is presented graphically in the zone rates plot. The y-scale can display density or holdups as relevant. Figure 3.32 illustrates a two-phase oil–gas situation with spinner and density. The dashed line represents the measured density. The current solution is such that for the selected correlation (Dukler), the predicted

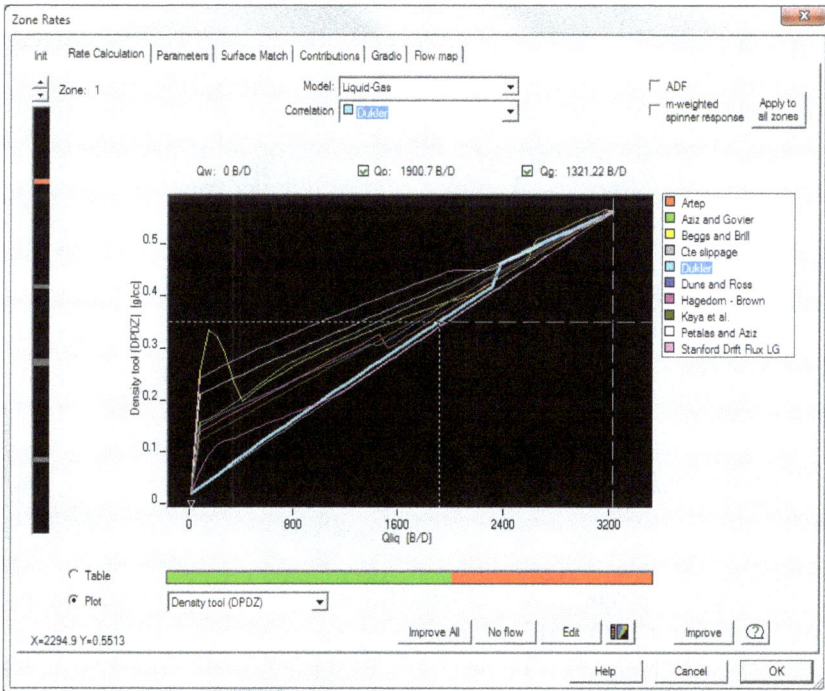

Figure 3.32. Emeraude zone rates dialogue.

density value (horizontal dotted line) is similar to the measured one (horizontal dashed line).

Each coloured line represents a different correlation. As we did previously, one way to look at this graph is to consider how different the rate solution will be depending on the selected correlation. The first step in a proper selection is to rule out the correlations developed for situations not applicable to the particular test. Beyond this first elimination, it is possible to check which correlation is the most consistent with additional information, surface rates in particular. This selection is a very important step of the analysis, and at least should be noted and justified. Software defaults will not be a sufficient excuse.

3.9.7. *Multi-Phase Interpretation Results*

Figure 3.33 illustrates a typical presentation. In this vertical oil–gas producer, three different (grey) calculation zones were selected to isolate the contributions of the two (red) perforations. The non-linear regression described previously was executed on all three calculation zones to get Q_o, and Q_g. Actually, on the bottom zone, the rate was set to 0 and water holdup to 1. On other zones, we have only two measurements (spinner and density) and we imposed $Q_w = 0$, so the regression solved only for Q_o and Q_g. The results are summarised by the QZT track where the values of a given calculation zone are extended up to the in-flow zone above, and down to the in-flow below.

The QZI track represents contributions or injections, and they are obtained basically by the difference of the rate above and below an in-flow. The Q track represents the application of the non-linear regression at every depth frame.

The two match views show the comparison between target (red) and simulated (green) measurements. The simulated curves are obtained by feeding at every depth the known rates into the forward model, including in particular the slip correlation. The simulated curves and the QZT logs are once again referred to as schematics.

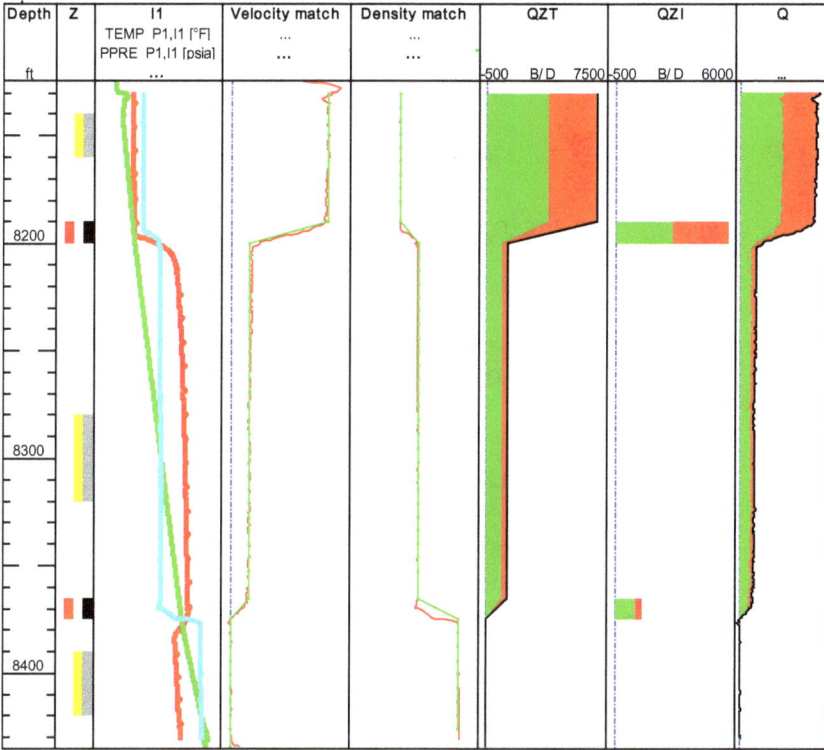

Figure 3.33. Typical multi-phase result; zoned approach.

3.9.8. *Local versus Global Regression*

The zoned approach described previously executes a series of unrelated non-linear regressions on the calculation zones; the contributions are then obtained by taking the difference of the rates above and below for each phase. By proceeding this way, there is no guarantee that in the end the contributions of a given zone will be of the same sign.

To avoid this, it is possible to solve for the entire well at once with a single regression, the global regression, where the unknowns are the zone contributions, dQs. Since the contributions are the direct unknowns, we can impose some sign constraints up front. For each iteration, the assumption of the dQs translates into a series of Qs on the calculation zones, which can be injected into the forward

			Local	Global
		$Q_w{}^1, Q_o{}^1, Q_g{}^1$	$E^1 = \Sigma\ (meas^1 - sim^1)^2$	$E = \Sigma\ (meas^1 - sim^1)^2$
$dQ_w{}^1, dQ_o{}^1, dQ_g{}^1$				+
		$Q_w{}^2, Q_o{}^2, Q_g{}^2$	$E^2 = \Sigma\ (meas^2 - sim^2)^2$	$\Sigma\ (meas^2 - sim^2)^2$
$dQ_w{}^2, dQ_o{}^2, dQ_g{}^2$				+
		$Q_w{}^3, Q_o{}^3, Q_g{}^3$	$E^3 = \Sigma\ (meas^3 - sim^3)^2$	$\Sigma\ (meas^3 - sim^3)^2$

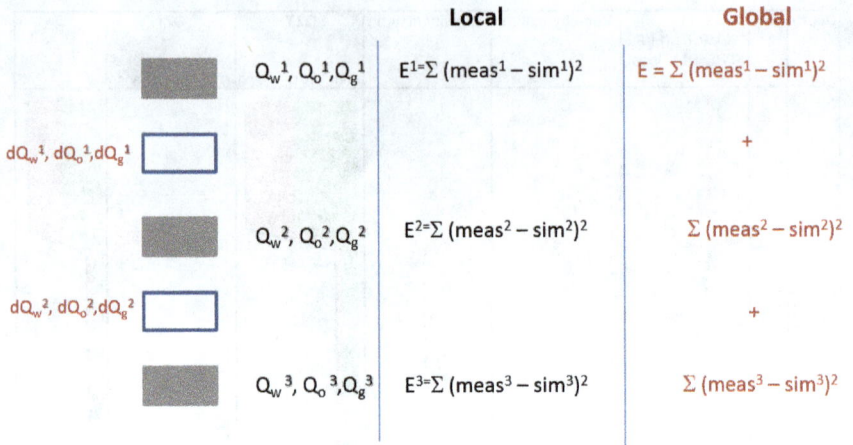

Figure 3.34. Local versus global regression.

model to calculate the objective function. Here again, the objective function is evaluated on the calculation zones only; to this error, other components might be added, such as a constraint using the surface rates.

Note that whether the regression is local or global, the end result is only influenced by the solution on the few user-defined calculation zones (Figure 3.34). This is why we call this approach the 'zoned' approach.

3.9.9.   *The Continuous Approach*

The clear advantage of the zoned approach is speed, since only a few points are required to get an answer, even if the global regression is finally run. Its main drawback is that the results are driven by the choice of calculation zones. A way to remove this dependency is to run a global regression with the errors evaluated *everywhere* on the logs, and not only at a few points. We could for instance seek to minimise the difference between the data and the simulated measurement logs (the schematics) everywhere. When we look at match views however, we see that the schematics are very square in shape; this is because between in-flow zones the mass rate does not change, and since we honour a slip model, there are little variations in holdups and deduced properties — see Figure 3.33 for instance.

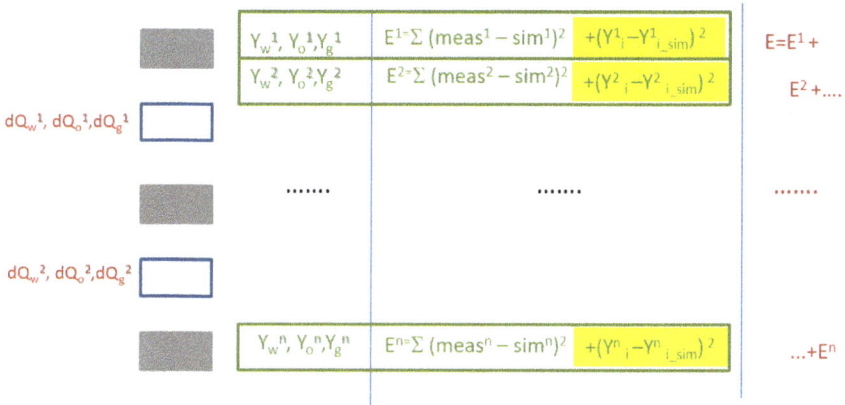

Figure 3.35. Global regression in continuous mode.

The only way to account for the changes seen in the data is to let the holdups differ from the model prediction locally, and at the same time to complement the objective function by a term measuring how far the solution deviates from the slip model prediction. More precisely, with the continuous approach, the global regression can be modified as follows (Figure 3.35).

The main regression loop is still on the contributions (red), but the objective function considers an error on the log points. In turn, at each depth, the simulated log values are evaluated by running a second regression on the holdups (green) to minimise an error made of the difference between simulated and measured values, and at the same time, a new constraint (yellow highlighting) using the slip model holdup predictions. In cases where one can do without the slip model, this new constraint is obviously not included (Figure 3.36).

3.9.9.1. *Illustration with a three-phase example*

Figure 3.36 shows the result of the zoned approach on a three-phase example. The calculation zones (grey) were defined — see how the second perforation has been split. Local regressions have been run, followed by a global regression with the constraint that all contributions are ≥ 0.

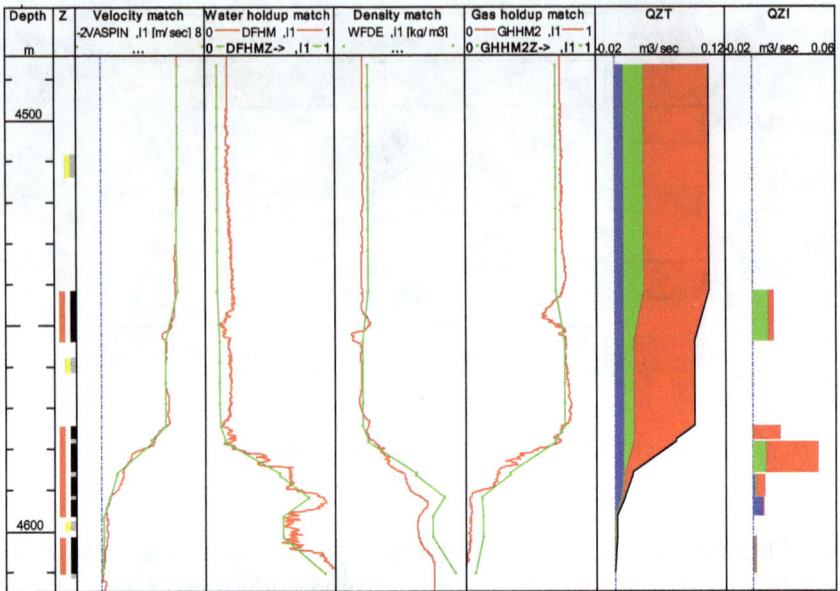

Figure 3.36. Example of zoned approach result.

This example is now interpreted with the continuous approach and the global regression rerun. In the end, there are certain differences. The match looks better overall, but at the expense of not honouring the slip models. The deviation from the slip models is indicated in the rightmost track (line = model, markers = solution). Note that the number of depth samples on the logs is user-defined (Figure 3.37).

3.9.9.2. *So?*

In the above example, there is no drastic benefit to using the continuous method, but there are cases where the reverse will apply, for instance when the data are unstable between contribution intervals. In this case, the location of the calculation zones may dangerously impact the zoned-method results. Note, however, that the continuous method is also influenced by the location of the calculation zones *inside the in-flows,* since the way the in-flows are split has a direct influence on the shape of the simulated logs over the in-flows. Another situation where the continuous method might

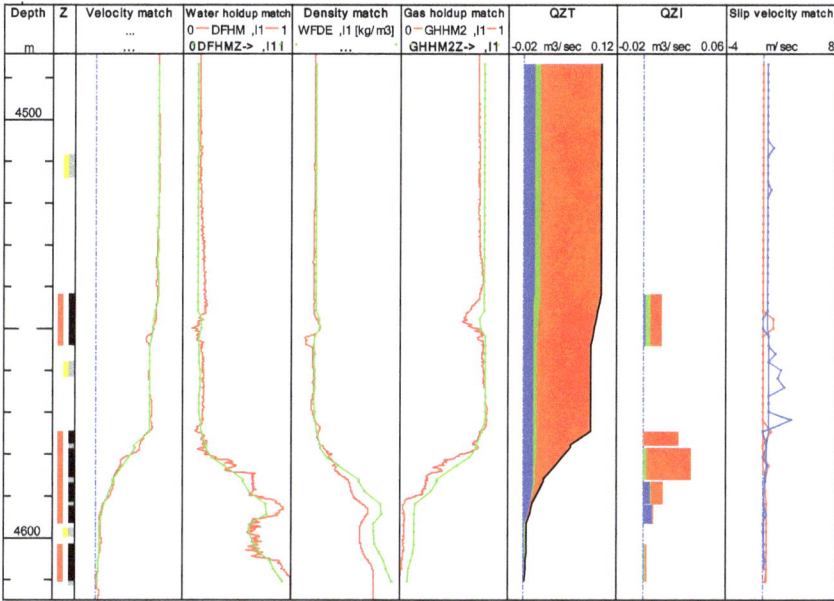

Figure 3.37. Example of continuous approach result.

provide a better answer is with temperature, since the temperature is essentially an integral response. The two methods, zoned and continuous, can be offered in parallel, to allow switching from one to the other at any stage.

It is important to stress that a nice-looking match does not necessarily mean a right answer. It all comes down at some stage to the interpreter's judgement. Also, any regression is biased by the weights assigned to the various components of the objective function. Different weights will lead to different answers; the starting point will also be critical, as a complex objective function will admit local minima. So, the continuous approach is not a magical answer, and more complex does not necessarily mean better. The continuous approach is more computing-intensive and a bit of a black box.

3.10. Slanted and Horizontal Wells

Even though the solutions presented up to now are in principle applicable to any well geometry, things get terribly complex in

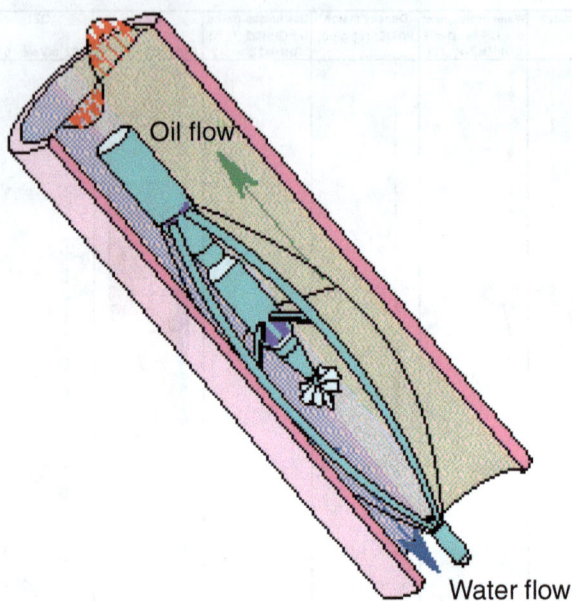

Figure 3.38. Flow recirculation.

deviated or horizontal wells. The response of a conventional tool in these environments can be unrepresentative of the flow behaviour; in addition, even if the tool responses can be trusted, the slippage models can turn out to be inadequate (Figure 3.38).

3.10.1.   *Apparent Down Flow*

In a slanted well with significant water holdup, the water phase tends to be circulated around while the light phase is going up. As the light phase occupies a small section of the pipe, the spinner will mostly see the water going down and a straight analysis will come up with a negative water rate. (The heavy phase is not limited to water and could also be oil.)

One way to deal with this situation is to use a dedicated model considering that the heavy phase is essentially static while the light phase moves up at a speed proportional to the slippage velocity. The only requirement is to have some holdup measurement and to deduce the slippage velocity by matching this holdup.

Another possible way of dealing with this situation is to use the temperature quantitatively as one of the target measurements. This obviously requires a forward temperature model representing the necessary thermal exchanges between the fluid, the reservoir and the well.

3.10.2. *Horizontal Wells*

The first problem in horizontal well logging is that the tools do not go down by gravity; dedicated conveyance systems are required, the two families being coiled tubing and tractors. The conveyance may affect the measurements, as for instance with tractors, when sometimes all the power goes into the tractor running in and one is only able to log coming up.

Horizontal wells are rarely strictly horizontal, and so unfortunately slippage velocities, and hence holdups, are very sensitive to slight changes of deviation around the horizontal (Figure 3.39).

The schematics in Figure 3.40, show this dependence. A 50–50 rate mixture is flown with water and oil. A blue dye is injected in the water and at the same time as a red dye in the oil. At 90°, both phases move at the same speed and the holdups are 50% each. At 88°, i.e. going uphill by only 2°, the oil flows much faster and its holdup decreases significantly. Conversely, going downhill by 2° at 92°, the situation is reversed with the water flowing faster.

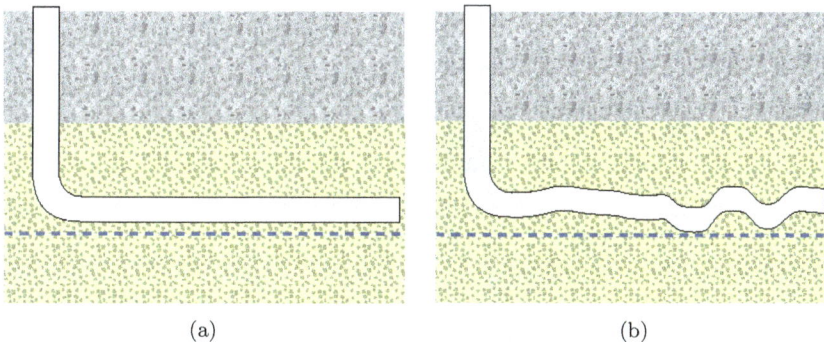

(a)   (b)

Figure 3.39. Horizontal well trajectory; ideal (a) and real (b).

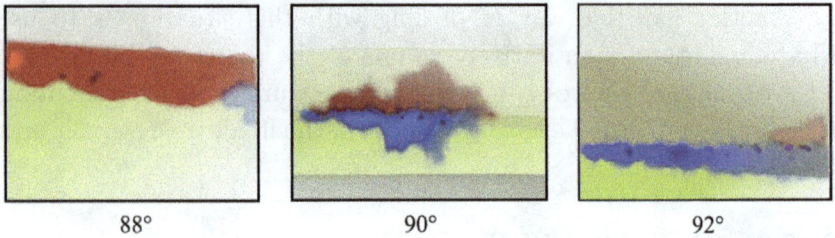

88°   90°   92°

Figure 3.40. Slippage dependence around horizontal.

It is not difficult to imagine that with undulations, the change from one behaviour to the other will not be immediate, leading to intermittent regimes such as waves. In this situation, the response of conventional tools will be useless most of the time. Even if they were reliable, slippage models that capture the physical behaviours in this situation are few.

The undulation of the wellbore will also create natural traps for heavy fluids in the lows and light fluid in the highs. Those trapped fluids will act as blocks and obviously impact the tool responses when they occur. A last condition worth mentioning is that the completion may offer flow paths not accessible to the conventional tool, e.g. with a slotted liner and multiple external packers.

For all the above reasons, specific tools were developed, called MPTs. The goal with those tools is to replace a single value response with a series of discrete points in order to better characterise the flow behaviour, and ultimately to remove the need for slippage models.

3.11. MPT, or Array Tools

3.11.1. *Schlumberger FloView*

The FloView is a generic name that includes the PFCS and the DEFT. The tools include four or six water holdup probes that use the electrical conductivity of water to distinguish between the presence of water and hydrocarbons.

In a water-continuous phase, current is emitted from the probe tip and returns to the tool body. A droplet of oil or gas has only to land on the probe tip to break the circuit and be registered.

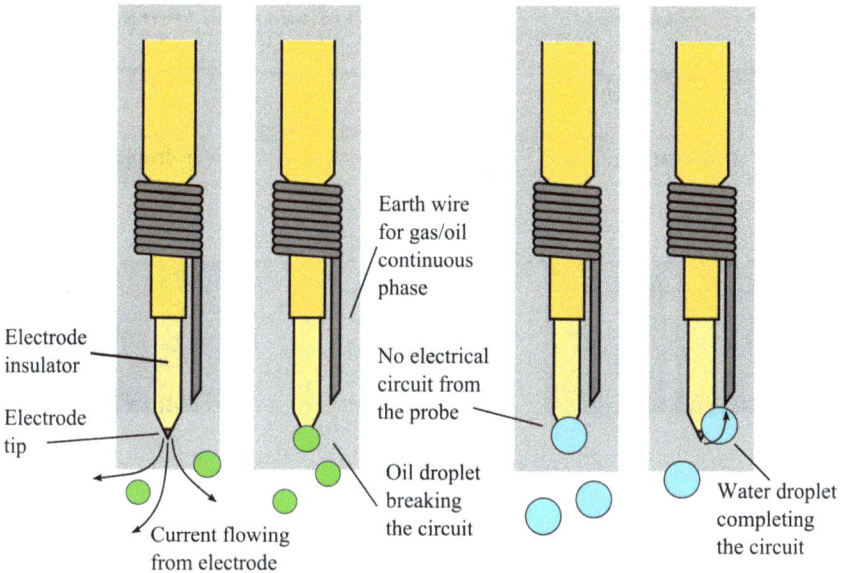

Figure 3.41. Theory of FloView probe operation.

In an oil-continuous phase a droplet of water touching the probe tip will not provide an electrical circuit. Instead, the water droplet must connect the electrical probe to the earth wire. This requires a larger droplet than is needed for gas or oil detection in a water-continuous phase (Figure 3.41).

The signal from the FloView probe lies between two baselines, the continuous water-phase response and the continuous hydrocarbon-phase response. To capture small transient bubble readings, a dynamic threshold is adjusted close to the continuous phase and then compared with the probe waveform. A binary water holdup signal results, which when averaged over time becomes the probe holdup. The number of times the waveform crosses the threshold is counted and divided by two to deliver a probe bubble count (Figure 3.42).

3.11.2. *Schlumberger GHOST*

The GHOST comprises four gas holdup probes. The probes use the refractive indices of gas, oil and water to distinguish between the presence of gas and liquid (Figure 3.43).

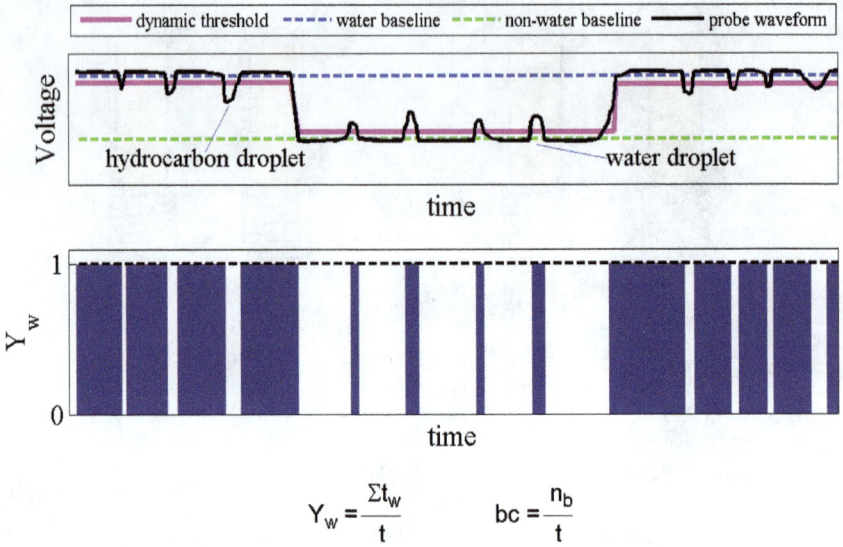

$$Y_w = \frac{\Sigma t_w}{t} \qquad bc = \frac{n_b}{t}$$

Figure 3.42. FloView probe waveform processing.

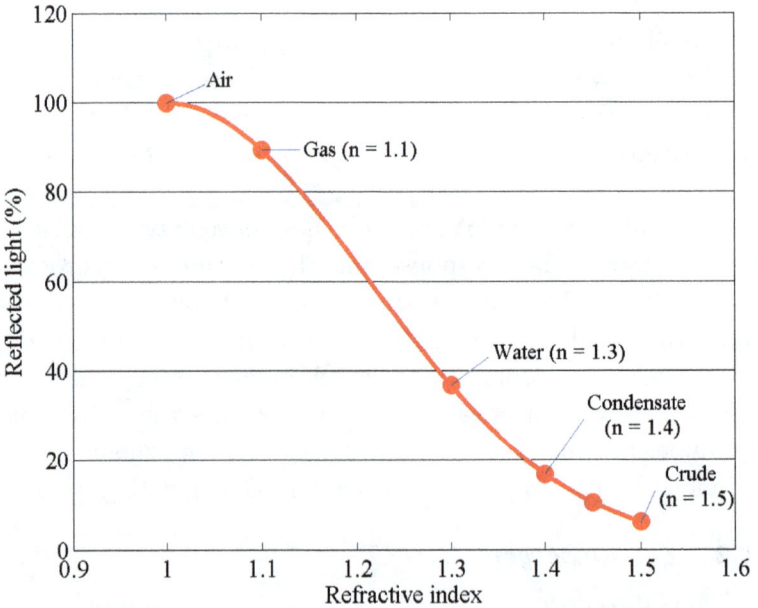

Figure 3.43. Reflected light versus refractive index.

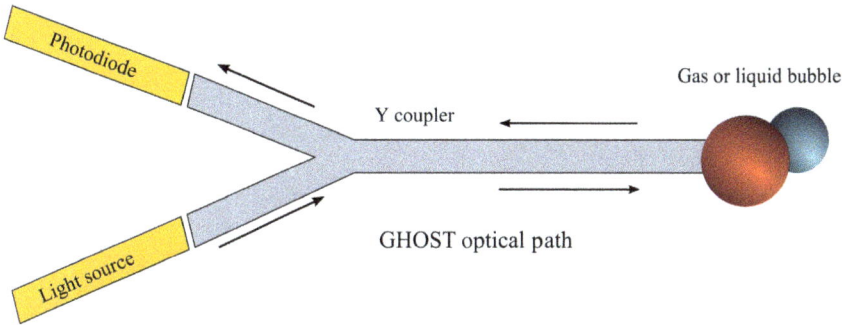

Figure 3.44. Sensor optical path.

Light emitted at a suitable frequency is fed down an optical fibre through a Y-coupler and finally to an optical probe made from synthetic sapphire crystal. Light that does not escape is returned via the Y-coupler to a photodiode and is converted to a voltage (Figure 3.44).

The signal from the optical probe is at or below the gas baseline and at or above the oil baseline. To capture small transient bubble readings, a dynamic threshold is adjusted close to the continuous gas phase and close to the continuous liquid phase. The threshold is then compared with the probe waveform to deliver a binary gas holdup signal, which is averaged over time. The number of times the waveform crosses the threshold is counted and divided by two to deliver a probe bubble count (Figure 3.45).

3.11.3. *Schlumberger FloScanner (FSI)*

The FSI combines five Micro-Spinners (MS) with six FloView probes (electrical — Ep) and six GHOST probes (optical — Op). The tool is designed to sit on the low side of the pipe by gravity, hence providing water holdup, gas holdup and velocity profiles on a vertical axis.

The operation principle of the holdup sensors is the one explained in the previous sections. The calibration of the individual FSI spinners is pretty straightforward, as the spinners are supposed to be in the same vertical location from one pass to the other.

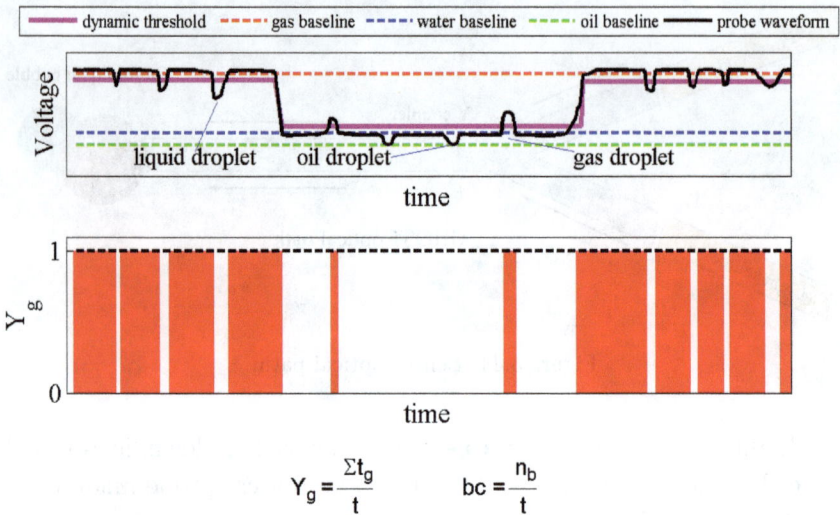

Figure 3.45. Optical probe waveform processing.

3.11.4.   *Sondex MAPS*: *CAT/RAT/SAT*

3.11.4.1.   *Capacitance Array Tool*

The Capacitance Array Tool (CAT) uses 12 capacitance probes distributed on a circumference. As with any capacitance sensor, the CAT probes discriminate mostly between water and hydrocarbons. The contrast in dielectric values for oil and gas may however be used to differentiate the two fluids.

In a stratified environment, the probe response can be used to get holdup values using two two-phase calibrations as represented in Figure 3.46. When the probe-normalised response is between 0 and 0.2, a gas–oil system is considered, and from 0.2 to 1, an oil–water system. To solve in three-phase without any assumption about the local holdups, a three-phase response can be used as shown in Figure 3.47. This response is an extension of the previous graph that constitutes the intersection of the surface with the side walls.

3.11.4.2.   *Resistance Array Tool*

The Resistance Array Tool (RAT) incorporates 12 sensors arranged on a circumference. The sensors mechanics comprise (1) a probe tip

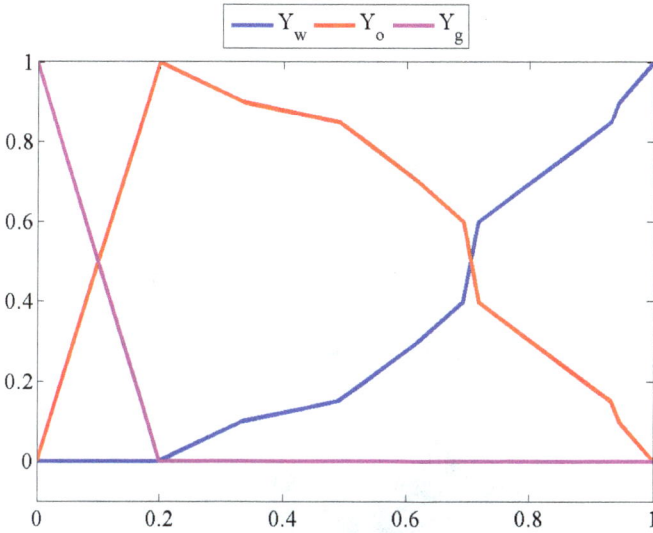

Figure 3.46. Normalised values in gas–oil then oil–water mixtures.

that ultimately connects to the sensor electronics input and (2) a reference contact, typically at earth potential.

Resistance measurements are made between the probe tips and result in a number that is proportional to the logarithm of the resistance detected between the electrodes, and therefore the fluid resistivity. The field outputs include a mean resistivity and a standard deviation over a small time window, as well as histograms. The usual processing is based on the mean resistivity, R, considered to be a linear function of the conductive water resistivity, R_c, and the insulating hydrocarbon resistivity, R_i.

The water holdup is thus obtained directly as

$$Y_w = \frac{R - R_i}{R_c - R_i}.\tag{3.19}$$

3.11.4.3. *Spinner Array Tool*

The Spinner Array Tool (SAT) uses six MS, again distributed on a circumference. One of the complexities of the SAT is that the MAPS tool string can usually freely rotate, and hence a SAT spinner will typically be at a different position in the pipe at a given depth

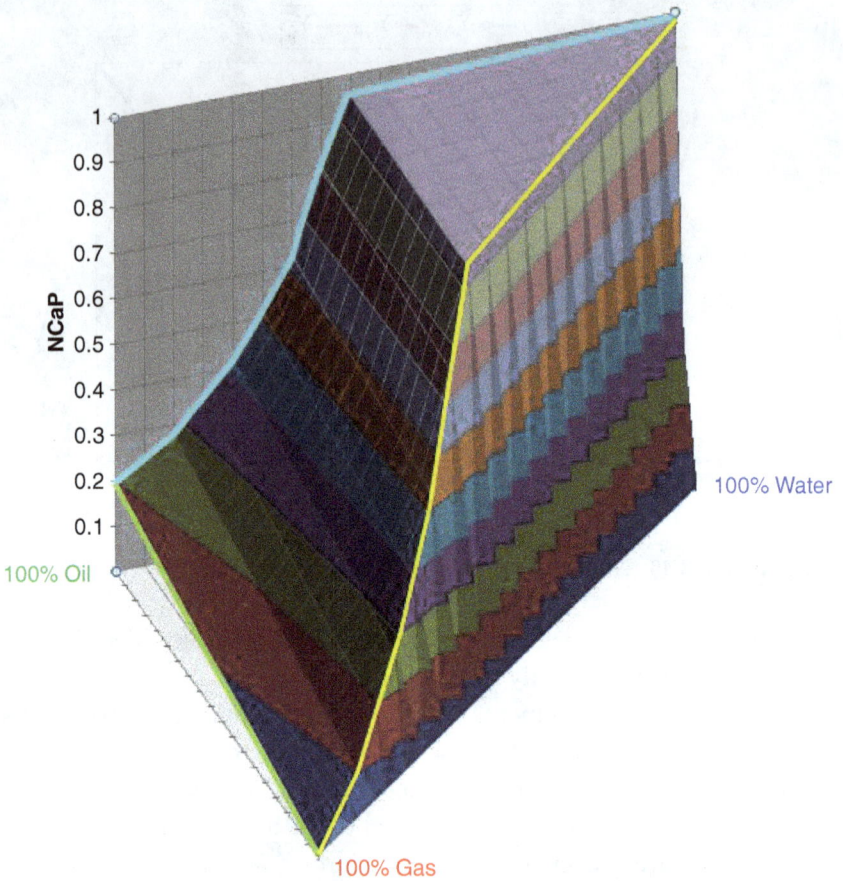

Figure 3.47. Three-phase CAT response. Note the two-phase segments:

Oil-Gas   Oil-Water   Gas-Water

for distinct passes. This makes the spinner calibration in flowing conditions problematic.

3.12.   MPT Interpretation

For any MPT tool, the exact position of the probes at a given depth is determined from the tool geometry, the local diameter and the tool bearing, which is part of the acquired measurements. The first required step of a quantitative analysis is to move from the discrete values to a Two-Dimensional (2D) representation. Having this 2D representation will then serve two purposes:

1. integration of the individual properties to give a representative average;
2. combination of the local properties to get phase rates and integration.

Imagine for instance that we know holdups and velocity everywhere. Locally, we can assume that there is no slippage, and at every location in the cross-section calculate the phase rates as the local velocity multiplied by the local holdup. By integrating the phase rates, we can get average phase rates directly, and therefore produce a final result without the need for slippage models.

3.12.1. *Mapping Models*

Mapping models assume horizontal stratification. If the flow regime is segregated, then a conventional analysis from the average holdups and velocity will be adequate.

3.12.1.1. *Linear model*

The linear model defines the measurement of interest by a number of variables, representing the values of that measurement along the local vertical axis. There are as many variables as there are distinct valid projections of probe readings on this axis. The values are then extended laterally. Without further constraints, the linear model will go exactly through the projected values. This is illustrated in Figure 3.48 with a RAT pass. The 12 projections define the values of water holdup on the vertical axis, and those values are extended laterally. The coloured 2D map of the water holdup (water = blue and oil = green) shows segregation, but the holdup does not strictly decrease from bottom to top.

To correct the previous situation, it is possible to alter the linear model using gravity segregation constraints, i.e. imposing that water holdup decreases from bottom to top, or gas holdup increases from bottom to top.

A non-linear regression can be used to try and match the values and satisfy the constraints at the same time. The result is shown in Figure 3.49.

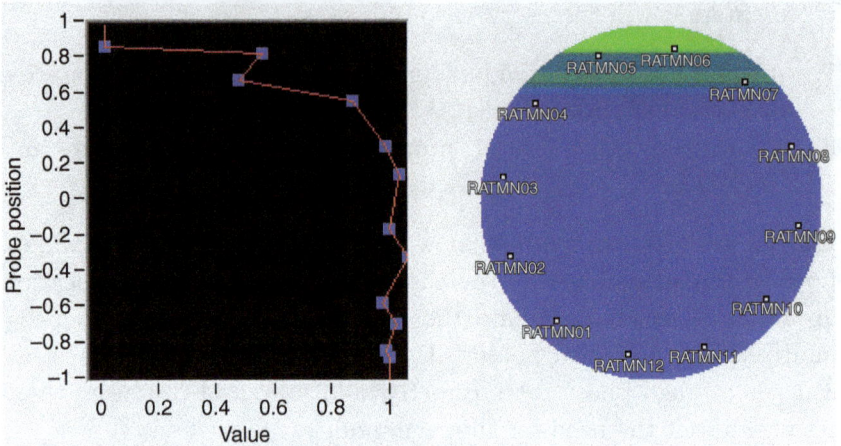

Figure 3.48. RAT mapping with the linear model; no constraint.

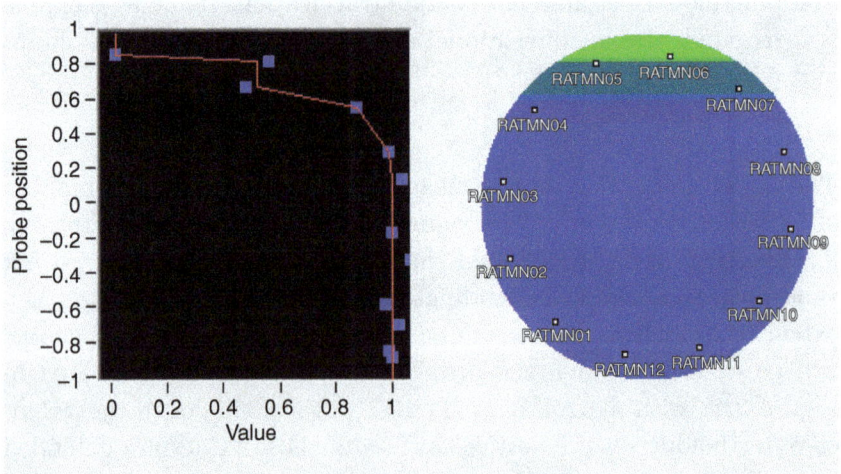

Figure 3.49. RAT mapping with the linear model; segregation constraint applied.

3.12.1.2.   *Schlumberger MapFlo holdup model*

This model is applicable to Schlumberger holdup measurements. It can be used with PFCS, DEFT, GHOST, FSI or even in combinations. All the model tries to match, like the linear model, is the projection of the measurements on the vertical axis. The MapFlo model is based on two parameters and produces the typical shapes/responses shown in Figure 3.50.

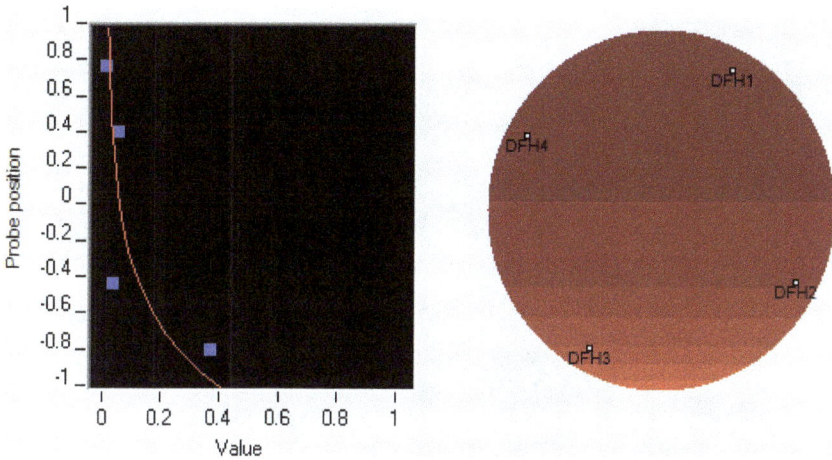

Figure 3.50. PFCS mapping with the MapFlo model.

3.12.1.3. *Prandtl velocity model*

This model can be used for the FSI velocity mapping. Like MapFlo, it is driven by two parameters. The main idea behind this model is to obtain the velocity profile by applying a linear transformation of the holdup profile and then rounding the profile near the pipe walls. More precisely, the velocity profile is obtained with Eq. (3.20), regressing on α and β to match the velocity projections:

$$[(Y_{\mathrm{w}} - Y_{\mathrm{g}}) \times \alpha + \beta] \times \left(1 - \left|\frac{z}{r}\right|\right)^{1/7}. \tag{3.20}$$

Figure 3.51 shows how the vertical holdup and velocity profiles were obtained on an FSI example with MapFlo and Prandtl combined.

The water holdup is 0 everywhere; the gas holdup profile is shown in red (see the Y_{g} scale at the bottom). The velocity profile is displayed with the yellow curve; one spinner was ignored in this case. The squares represent the discrete measurements (blue $= Y_{\mathrm{w}}$, red $= Y_{\mathrm{g}}$, yellow $= V$).

It should be noted that the Prandtl model rounds the edge of the velocity on the entire circumference and not just at the top and bottom.

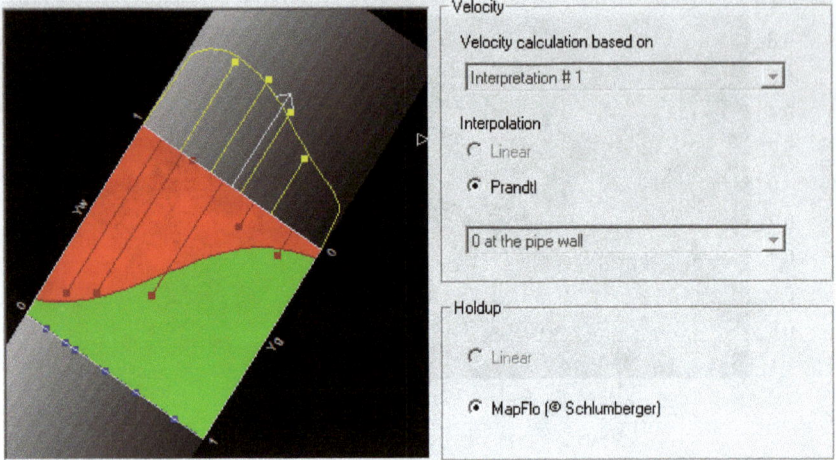

Figure 3.51. FSI mapping in a deviated well with Mapflo and Prandtl.

3.12.2. *Mapping Options*

We saw how the basic framework used to map the 2D models is based on non-linear regression. This framework offers a lot of flexibility in the number of inputs as well as the inclusion of external constraints. We mentioned the ability to associate gravity segregation constraints with the linear model. It is also possible to constrain the process by the value of a conventional tool, e.g. a density tool for instance.

Finally, the optimisation can be based on several passes at the same time if the conditions are stable from one pass to the other. When the tool rotates (MAPS, PFCS, GHOST), this provides the ability to multiply the points of measurements in the cross-section at every depth and can compensate for faulty probes.

3.12.3. *Integration*

The mapping allows integrating over the cross-section at every depth in order to get average values:

$$Y_{\mathrm{w}} = \frac{\int_S Y_{\mathrm{w}} \times dS}{S}; \quad Y_{\mathrm{g}} = \frac{\int_S Y_{\mathrm{g}} \times dS}{S};$$

$$Y_{\mathrm{o}} = \frac{\int_S Y_{\mathrm{o}} \times dS}{S}; \quad V = \frac{\int_S V \times dS}{S}. \tag{3.21}$$

As explained earlier, with the assumption of no local slippage, we can use the local velocity and holdups to obtain phase rates:

$$Q_w = \frac{\int_S Y_w \times V \times dS}{S}; \quad Q_o = \frac{\int_S Y_o \times V \times dS}{S};$$

$$Q_g = \frac{\int_S Y_g \times V \times dS}{S}. \tag{3.22}$$

3.12.4. *Interpretation*

The interpretation is conducted using the process outputs: holdups, phase rates and total velocity, together with any additional tool available. Even though the answers we are seeking are in essence already provided (we have the phase rates everywhere), the interpretation is still a required step to come up with actual zone contributions, possibly honouring additional constraints (sign, surface rates).

Using the continuous method is the best choice as it contains a built-in mechanism to bypass any slippage model on the recognition that enough information is supplied. Figure 3.52 is a typical output example for an FSI job.

3.13. SIP

SIP provides a means of establishing the steady state inflow relationship for each producing layer.

The well is flowed at several different stabilised surface rates, and for each rate a production log is run across the entire producing interval to record simultaneous profiles of downhole flow rates and flowing pressure. Measured *in situ* rates can be converted to surface conditions using PVT data.

The SIP theory is strictly valid for single-phase.

For each survey/interpretation, a couple (rate, pressure) corresponds to each reservoir zone used in the SIP calculation. For the pressure, the interpretation reference channel is interpolated at the top of the zone. For the rate, the value used in the SIP is the contribution. It is calculated for a given reservoir zone as the difference between the values interpolated on the schematic at the top and the bottom of that zone (Figure 3.53).

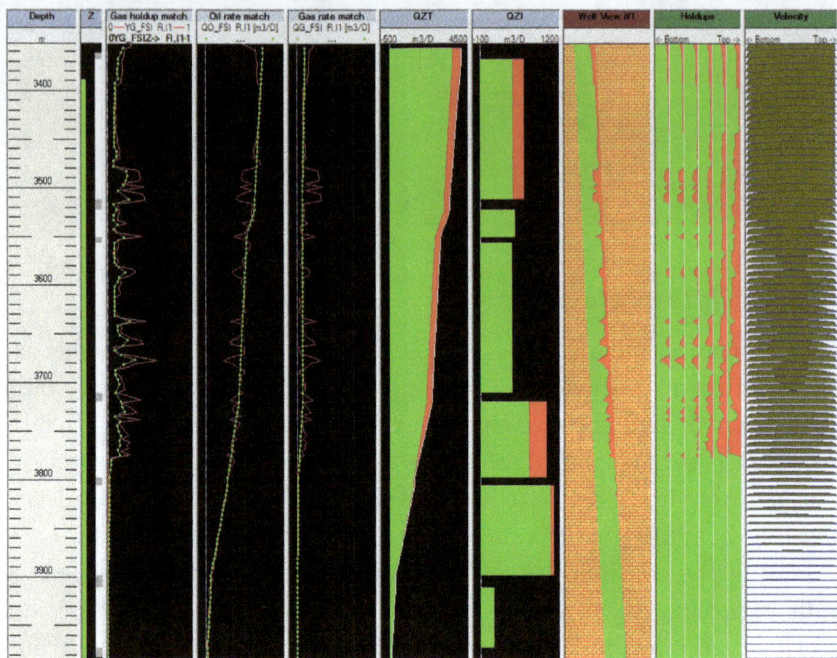

Figure 3.52. FSI Interpretation.

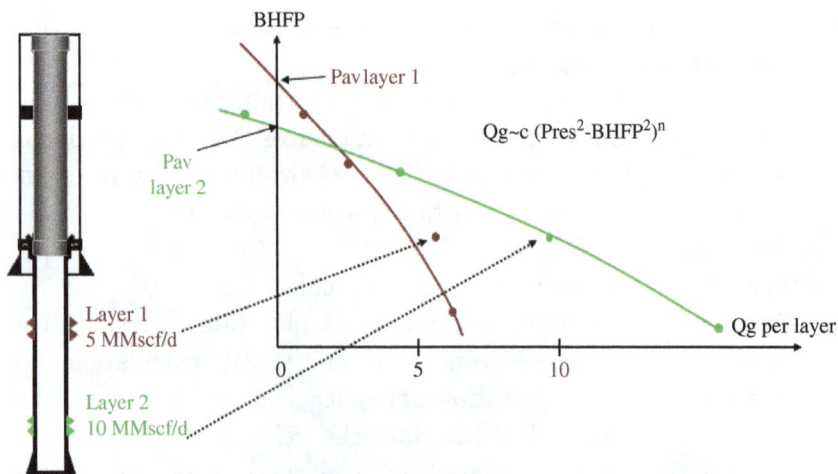

$$Qg \sim c\,(Pres^2 - BHFP^2)^n$$

Figure 3.53. SIP example with two layers, three rates and a shut-in survey.

3.13.1. *IPR Type*

Different Inflow Performance Relation (IPR) equations can be used: straight line, C&n and LIT relations. In the case of a gas well, the pseudo pressure, $m(p)$, can be used instead of the pressure, p, to estimate the gas potential.

Straight line:

$$Q = PI \times (\bar{p} - p), \tag{3.23}$$

LIT (a&b):

$$\bar{p}^2 - p^2 = a \times Q + b \times Q^2, \tag{3.24}$$

or

$$m(\bar{p}) - m(p) = a \times Q + b \times Q^2, \tag{3.25}$$

Fetkovitch or C&n:

$$Q = C \times (\bar{p}^2 - p^2)^n, \tag{3.26}$$

or

$$Q = C \times (m(\bar{p}) - m(p))^n. \tag{3.27}$$

It is possible to generate the SIP after correcting the pressures to a common datum. This is typically done using the shut-in pressure profile for the estimation of the hydrostatic head between layers. A pressure-corrected SIP highlights the eventual cross-flow between layers (Figure 3.54).

3.14. Temperature

Temperature can be used quantitatively to replace a missing or faulty spinner, provided an adequate temperature model is available. This model should provide the temperature everywhere in the wellbore,

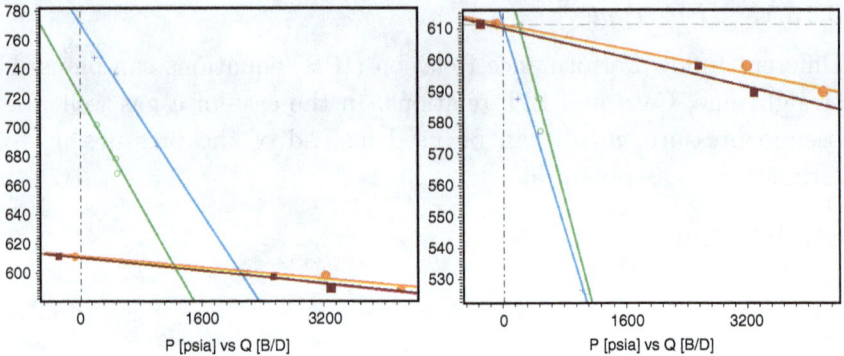

Figure 3.54. SIP example without (left) and with (right) datum correction.

from an assumed distribution of contributions. Such a model needs to capture the following:

- Temperature changes occurring inside the reservoir (compressible effects, friction). These are often reduced to simply a Joule–Thomson cooling/heating, but the reality is more complex.
- Temperature changes in the wellbore — due to convection (transport), conduction to the surrounding, changes of enthalpy in front of inflows, and changes due compressible effects in the wellbore.

It is beyond the scope of this chapter to describe temperature models in detail. They can be sometimes be very simple, or capture all the above effects by solving numerically a general energy equation coupled to a mass balance equation. Figure 3.55 shows an example match in an apparent downflow situation.

Working with temperature requires a large number of inputs, typically geothermal profile, rock thermal properties, fluid heat capacities, thermal properties of the completion elements, reservoir petrophysical properties, etc. If these parameters are not available, they become additional degrees of the freedom of the problem, leading to multiple answers. In any case, the discrimination of the phases still needs fluid identification measurements like density or holdups.

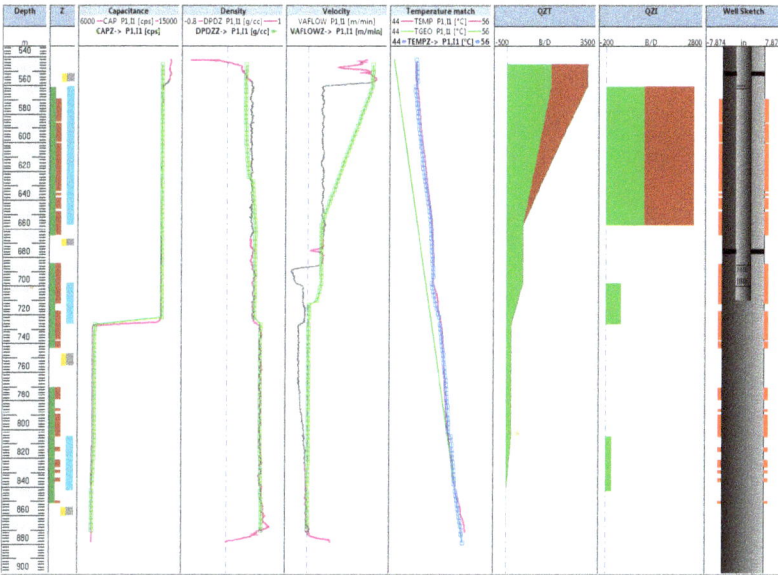

Figure 3.55. Temperature match in Emeraude.

References

[1] Hill, A.D. (1990). *Production Logging, SPE Monograph Series*, Vol. 14. Society of Petroleum.

[2] Whittaker, C.C. (2013). *Fundamentals of Production Logging*, Schlumberger.

Index

www.ingramcontent.com/pod-product-compliance
Lightning Source LLC
Chambersburg PA
CBHW072256210326
41458CB00074B/1801